Intelligent Systems Reference Library

Volume 106

Series editors

Janusz Kacprzyk, Polish Academy of Sciences, Warsaw, Poland
e-mail: kacprzyk@ibspan.waw.pl

Lakhmi C. Jain, Bournemouth University, Fern Barrow, Poole, UK, and
University of Canberra, Canberra, Australia
e-mail: jainlc2002@yahoo.co.uk

About this Series

The aim of this series is to publish a Reference Library, including novel advances and developments in all aspects of Intelligent Systems in an easily accessible and well structured form. The series includes reference works, handbooks, compendia, textbooks, well-structured monographs, dictionaries, and encyclopedias. It contains well integrated knowledge and current information in the field of Intelligent Systems. The series covers the theory, applications, and design methods of Intelligent Systems. Virtually all disciplines such as engineering, computer science, avionics, business, e-commerce, environment, healthcare, physics and life science are included.

More information about this series at http://www.springer.com/series/8578

Anna Esposito · Lakhmi C. Jain
Editors

Toward Robotic Socially Believable Behaving Systems - Volume II

Modeling Social Signals

 Springer

Editors
Anna Esposito
Department of Psychology
Seconda Università di Napoli and IIASS
Caserta
Italy

Lakhmi C. Jain
Faculty of Science and Technology
Data Science Institute, Bournemouth
  University
Fern Barrow, Poole
UK

ISSN 1868-4394   ISSN 1868-4408   (electronic)
Intelligent Systems Reference Library
ISBN 978-3-319-80950-2   ISBN 978-3-319-31053-4   (eBook)
DOI 10.1007/978-3-319-31053-4

This Springer imprint is published by Springer Nature
The registered company is Springer International Publishing AG Switzerland

Preface

When it comes to modeling emotions, contextual instances cannot be neglected. The concept of context is, to a certain extent, a complex one, since it includes cultural, social, physical, and individual features that shape human interactional exchanges. This second volume accounts for contexts, in particular, social contexts and social signals that must be interpreted to correctly and successfully decode the semantic and emotional meaning of interactional exchanges. To this aim, several experts, from different scientific domains, are describing behaviors to be adopted or interpreted for, as well as mathematical algorithms to model contextual instances and relative information communication technology (ICT) interfaces for developing robotic socially believable applications. In this regard, the volume presents the recent research works on robotics approaching the domestic spheres and recent research efforts for allowing robotic systems of automaton levels of intelligence, where "intelligent" is the system's ability to implement a natural interaction with human.

The implementation of such context-aware situated ICT systems should contribute to improve the quality of life of the end users through: (1) The development of shared digital data repositories and annotation standards for benchmarking; (2) new methods for data processing and data flow coordination through synchronization, temporal organization, and optimization of new encoding features (identified through human behavioral analyses); and (3) computational models synthesizing the human ability to rule individual choices, perception, and actions. The final goal would be to produce machines equipped with human-level automaton intelligence.

The editors would like to thank the contributors and the International Scientific Committee of reviewers listed below for their rigorous and invaluable scientific revisions, dedication, and priceless selection process. Thanks are also due to the Springer-Verlag for their excellent support during the development phase of this research book.

Italy   Anna Esposito
Australia   Lakhmi C. Jain

International Scientific Committee

- Samantha Adams (Plymouth University, UK)
- Samer Al Moubayed (School of Computer Science and Communication, Stockholm, Sweden)
- Ivana Baldassarre (Seconda Università di Napoli, Italy)
- Tony Belpaeme (Plymouth University, UK)
- Štefan Beňuš (Constantine the Philosopher University, Nitra, Slovakia)
- Ronald Böck (Otto von Guericke University Magdeburg, Germany)
- Branislav Borova (University of Novi Sad, Serbia)
- Nikolaos Bourbakis (Wright State University, Dayton, USA)
- Angelo Cafaro (Telecom ParisTech, Paris, France)
- Angelo Cangelosi (Plymouth University, UK)
- Chloe Clavel (Telecom ParisTech, Paris, France)
- Gennaro Cordasco (Seconda Università di Napoli and IIASS, Italy)
- Conceição Cunha (Ludwig-Maximilians University of Munich, Germany)
- Alessandro Di Nuovo (Plymouth University, UK)
- Thomas Drugman (University of Mons, Belgium)
- Stéphane Dupont (University of Mons, Belgium)
- Anna Esposito (Seconda Università di Napoli and IIASS, Italy)
- Antonietta Maria Esposito (Osservatorio Vesuviano, Napoli, Italy)
- Marcos Faúndez-Zanuy (EUPMt, Barcelona, Spain)
- Maria Giagkou (Institute for Language and Speech Processing, Greece)
- Milan Gnjatovič (University of Novi Sad, Serbia)
- Maurice Grenberg (New Bulgarian University, Sofia, Bulgaria)
- Jonathan Harrington (Ludwig-Maximilians University of Munich, Germany)
- Jennifer Hofmann (University of Zurich, Switzerland)
- Phil Hole (Ludwig-Maximilians University of Munich, Germany)
- Evgeniya Hristova (New Bulgarian University, Sofia, Bulgaria)
- Lazslo Hunyadi (University of Debrwcen, Hungary)
- Randy Klaassen (University of Twente, The Netherlands)
- Maria Koutsombogera (Institute for Language and Speech Processing, Greece)

- Barbara Lewandowska-Tomaszczyk (University of Lodz, Poland)
- Karmele Lopez-De-Ipina (Basque Country University, Spain)
- Saturnino Luz (University of Edinburgh, UK)
- Mauro Maldonato (Università della Basilicata, Italy)
- Rytis Maskeliunas (Kaunas University of Technology, Lithuania)
- Olimpia Matarazzo (Seconda Università di Napoli, Italy)
- Jiří Mekyska (Brno University of Technology, Czech Republic)
- Francesco Carlo Morabito, (Università "Mediterranea" di Reggio Calabria, Italy)
- Hossein Mousavi (Istituto Italiano di Tecnologia, Genova, Italy)
- Vittorio Murino (Istituto Italiano di Tecnologia, Genova, Italy)
- Costanza Navarretta (Centre for Language Technology, Njalsgade, Denmark)
- Rieks Op Den Akker (University of Twente, The Netherlands)
- Harris Papageorgiou (Institute for Language and Speech Processing, Greece)
- Eros Pasero (Politecnico di Torino, Italy)
- Jiří Přibil (Academy of Sciences, Czech Republic)
- Anna Přibilová (Slovak University of Technology, Slovakia)
- Zófia Ruttkay (Moholy-Nagy University of Art and Design Budapest, Hungary)
- Matej Rojc (University of Maribor, Slovenia)
- Michele Scarpiniti (Università di Roma "La Sapienza", Italy)
- Filomena Scibelli (Seconda Università di Napoli, Italy)
- Björn W. Schuller (Imperial College, UK and University of Passau, Germany)
- Zdenek Smékal (Brno University of Technology, Czech Republic)
- Jordi Solé-Casals (University of Vic, Spain)
- Stefano Squartini (Università Politecnica delle Marche, Italy)
- Igor Stankovic (University of Gothenburg, Sweden)
- Jing Su (Trinity College Dublin, Ireland)
- Jianhua Tao (Chinese Academy of Sciences, P.R. China)
- Alda Troncone (Seconda Università di Napoli and IIASS, Italy)
- Alessandro Vinciarelli (University of Glasgow, UK)
- Carl Vogel (Trinity College of Dublin, Ireland)
- Jerneja Žganec Gros (Alpineon, Development and Research, Slovenia)
- Paul A. Wilson (University of Lodz, Poland)
- B. Yegnanarayana (International Institute of Information Technology, Gachibowli, India)

Sponsoring Organizations

- Seconda Università di Napoli, Dipartimento di Psicologia
- International Institute for Advanced Scientific Studies "E.R. Caianiello" (IIASS, www.iiassvietri.it/), Italy
- Società Italiana Reti Neuroniche (SIREN, www.associazionesiren.org/)

Contents

About the Editors

Anna Esposito received her "Laurea Degree" *summa cum laude* in Information Technology and Computer Science from the Università di Salerno in 1989 with a thesis on: *The Behavior and Learning of a Deterministic Neural Net* (published on **Complex System**, vol 6(6), 507–517, **1992**). She received her PhD Degree in Applied Mathematics and Computer Science from the Università di Napoli, "Federico II" in 1995. Her PhD thesis was on: *Vowel Height and Consonantal Voicing Effects: Data from Italian* (published on **Phonetica**, vol 59(4), 197–231, **2002**) at Massachusetts Institute of Technology (MIT), Research Laboratory of Electronics (RLE), under the supervision of professor Kenneth N. Stevens.

She has been a Post Doc at the International Institute for Advanced Scientific Studies (IIASS), and Assistant Professor at the Department of Physics at the Università di Salerno (Italy), where she taught courses on Cybernetics, Neural Networks, and Speech Processing (1996–2000). She had a position as Research Professor (2000–2002) at the Department of Computer Science and Engineering at Wright State University (WSU), Dayton, OH, USA. She is currently associated with WSU as Research Affiliate.

Anna is currently working as an Associate Professor in Computer Science at the Department of Psychology, Seconda Università di Napoli (SUN). Her teaching responsibilities include Cognitive and Algorithmic Issues of Multimodal Communication, Human Machine Interaction, Cognitive Economy, and Decision Making. She authored 160+ peer reviewed publications in international journals, books and conference proceedings. She edited/co-edited 21 books and conference proceedings with Italian, EU and overseas colleagues.

Anna has been the Italian Management Committee Member of:

- COST 277: Nonlinear Speech Processing, http://www.cost.esf.org/domains_ actions/ict/Actions/277 (2001–2005)
- COST MUMIA: **MUltilingual and Multifaceted Interactive information Access**, www.cost.esf.org/domains_actions/ict/Actions/IC1002 (2010–2014)
- COST TIMELY: Time in Mental Activity, www.timely-cost.eu (2010–2014)

She has been the proposer and chair of COST 2102: Cross Modal Analysis of Verbal and Nonverbal Communication, http://www.cost.esf.org/domains_actions/ ict/Actions/2102 (2006–2010).

Since 2006, she is a Member of the **European Network for the Advancement of Artificial Cognitive Systems, Interaction and Robotics** (www.eucognition.org);

She is currently the Italian Management Committee Member of ISCH COST Action IS1406: Enhancing children's oral language skills across Europe and beyond (http://www.cost.eu/COST_Actions/isch/Actions/IS1406).

Anna's research activities are on the following three principal lines of investigations:

- **1998 to date**: Cross-modal analysis of speech, gesture, facial and vocal expressions of emotions. Timing perception in language tasks.
- **1995 to date**: Emotional and social believable Human-Computer Interaction (HCI).
- **1989 to date**: Neural Networks: learning algorithm, models and applications

Lakhmi C. Jain serves as a Visiting Professor in Bournemouth University, United Kingdom and Adjunct Professor in the Faculty of Education, Science, Technology and Mathematics in the University of Canberra, Australia. He is a Fellow of the Institution of Engineers Australia.

Dr. Jain founded the KES International for providing a professional community the opportunities for publications, knowledge exchange, cooperation and teaming. Involving around 5000 researchers drawn from universities and companies world-wide, KES facilitates international cooperation and generate synergy in teaching and research. KES regularly provides networking opportunities for professional community through one of the largest conferences of its kind in the area of KES.

www.kesinternational.org

His interests focus on the artificial intelligence paradigms and their applications in complex systems, security, e-education, e-healthcare, unmanned air vehicles and intelligent systems.

Chapter 1
Moving Robots from Industrial Sectors to Domestic Spheres: A Foreword

Leopoldina Fortunati

Abstract This foreword analyses how in the shift from industrial sectors to social and domestic sectors the form of robots is actually subjected to great changes. Furthermore, it reports some recent data on robot diffusion, implicitly inviting researchers to reflect on the trends of robot diffusion that give the precise indication to focus on the material sphere of the housework.

Keywords Social robotic autonomous systems · Physical · Social and organizational context · Social robots

This volume represents an important step for social robotics studies. It shows how this field of knowledge is advancing consistently on making increasingly "natural" human interactional exchanges with social robotic autonomous systems. Said this, the present collection does not hide the limitations of the research carried out so far on modeling social signals and calls for a strong engagement by social sciences. Implicating also social sciences is in fact the only way for assuring the improvement of the quality of life of the end-users of social robots [1–5].

The focus of this volume is on the physical, social and organizational context, which is proved to be so crucial in affecting the whole communication and interaction process. The main characteristics of human interaction exchanges are explored and the cognitive and emotional processes at the core of interactions are modeled. The purpose behind this research is to develop robotic system prototypes able to socially behave in specific contexts. The research on social robotics field needs to advance substantially because now many different care sectors, the healthcare, the entertainment, the education, and the domestic setting are pressing for useful products.

The shift from the industrial setting to the reproduction sphere of individuals obliges us to be aware that it intertwines in parallel with another shift from the unitary conception of robots as machines to a more articulated conception of robots.

L. Fortunati (✉)
Dipartimento di Studi Umanistici e del Patrimonio Culturale,
Università di Udine, Pordenone, Italy
e-mail: leopoldina.fortunati@uniud.it

© Springer International Publishing Switzerland 2016
A. Esposito and L.C. Jain (eds.), *Toward Robotic Socially Believable
Behaving Systems - Volume II*, Intelligent Systems Reference Library 106,
DOI 10.1007/978-3-319-31053-4_1

We inherited from the past some forms of robot: zoomorphic; android or gynoid; machine with a very different profile from human beings; particular objects [6]. However, with the current proliferation of technologies in the social body we do not need now to remain bound to these forms. What is increasingly emerging is the attempt to take advantage of the network of information and communication technologies so diffused in the social body for developing robotic systems [7]. Thus, along with the development of robots as continuation of the classic forms of robot—agents as drones, humanoids, swarms, cleaning, manufacturing, emergency and space exploration robots-, we face in fact the de-construction of robots in many new forms: from intelligent agents to automated personal assistants, from future smart environments to ambient assistive living technologies, from computational intelligent games/storytelling devices to embodied conversational avatars, and automatic healthcare and education services [8].

This shift is made possible by the new level reached by social robotics science, of which this volume is an important witness, and is made necessary by two factors. First, the new social needs and desires expressed by people at immaterial level [9, 10], which call for practical and innovative solutions and can be addressed through robots or algorithms. I allude here to the increasing use of robots in the Web 2.0, in information, education, entertainment, health and care sectors, which can be considered as forms of proto-roboticization of the immaterial sphere in society.

Second, the domestic needs at material level call also to be addressed: see the extraordinary success of vacuum cleaner and floor cleaner robots and the launch of several kitchen robot (which are multi-functional food processor, cooker, steamer and self-cleaner) that propose to radically change the way we cook, learn about cooking, coordinate and time plan our food practices at home.

To have a more precise idea of the diffusion of social robots, it is wise to recur to the World Robotics Report 2015, released by the Statistical Department of the International Federation of Robotics (IFR) [11]. In this report industrial and service robots (those sold for professional or personal/domestic use) are distinguished. Social robots are part of the service robots. According to this Report, in 2014 professional service robot sales arrived to 24,207 units, showing an increase of 11.5 % compared to 2013. But service robots for personal and domestic use arrived to 4.7 million units, showing an increase of 28 % compared to 2013. Disabled assistance robots have taken off in 2014, when a total of 4,416 robots were sold with an increase of 542 % in respect to 2013 when only 699 units were sold.

However, the bulk of sales occurred in the sector of robots for domestic tasks: 3.3 million of units were sold, including vacuum cleaning, lawn-mowing, window cleaning and other types of. In the entertainment sector, there are increasingly more sophisticated low-priced toy robots, but also high quality products such as those proposed by the LEGO® Mindstorms® programme, offering software environments that reach actually into high-tech robotics. These digits tell us that the domestic material needs are those more urgent to respond to and that there is a huge space for market development in the domestic sphere. Also the projections for the period 2015–2018 reported by IFR show that 35 million of service robots for personal use are expected to be sold. By contrast, the sales of robot companions/assistants/humanoids are projected to be only 8,100 units in the same period. In reality, these social robots started to arrive to research laboratories and university only in 2004, but so far there have not been significant sales of robots as human companions to perform typical everyday tasks.

All these processes converge on bringing to light what I call the new phenomenon of "ubiquitous social roboting". The new researches on social robotics, such as those presented and discussed in this volume, are timely and necessary to strengthen this phenomenon by developing further and in the right direction (from a human perspective) the robotification of society [12].

References

1. Sugyiama S, Vincent J (eds) (2013) Social robots and emotion: transcending the boundary between humans and ICTs. Intervalla: Platform for Intellectual Exchange 1(Special issue)
2. Fortunati L, Esposito A, Ferrin G, Viel M (2014) Approaching social robots through playfulness and doing-it-yourself: children in action. Cogn Comput 6(4):789–801
3. Fortunati L, Esposito A, Sarrica M, Ferrin G (2015) Children's knowledge and imaginary about robots. Int J Soc Robot 7(5):685–695
4. Höeflich, JR, El Bayed A (2015) Perception and acceptance of social robots - exploratory studies. In: Vincent J, Taipale S, Sapio B, Lugano G, Fortunati L (eds) Social robots from a human perspective. Springer, London, pp 39–54
5. Katz JE, Halpern D, Thomas Crocker E (2015) In the company of robots: views of acceptability of robots in social settings. In: Vincent J, Taipale S, Sapio B, Lugano G, Fortunati L (eds) Social robots from a human perspective. Springer, London, pp 25–38
6. Fortunati L (2013) Afterword: robot conceptualizations between continuity and innovation. Intervalla: Platform for Intellectual Exchange, vol 1. pp 116–129
7. Fortunati L, Esposito A, Lugano G (2015) Beyond industrial robotics: social robots entering public and domestic spheres. Inf Soc: Int J 31(3):229–236
8. Esposito A, Fortunati L, Lugano G (2014) Modeling emotion, behaviour and context in socially believable robots and ICT interfaces. Cogn Comput 6(4):623–627
9. Fortunati L (1981) L'arcano della riproduzione: casalinghe, prostitute, operai e capitale. Venezia: Marsilio (The arcane of reproduction: housework, prostitution, labor and capital). Autonomedia, New York 1995
10. Fortunati L (2007) Immaterial labor and its machinization. Ephemer Theory Polit Organ 7(1):139–157
11. IFR Statistical Department, World Robotics Report 2015. http://www.ifr.org
12. Vincent J, Taipale S, Sapio B, Lugano G, Fortunati L (eds) (2015) Social robots from a human perspective. Springer, London

Chapter 2
Modeling Social Signals and Contexts in Robotic Socially Believable Behaving Systems

Anna Esposito and Lakhmi C. Jain

Abstract There is a need for a holistic perspective when considering aspects of natural interactions with robotic socially believable behaving systems, that must account of the cultural, social, physical, and individual (the context) features that shape interactional exchanges. Context (the physical, social and organizational context) rules individual's social conducts and provide means to render the world sensible and interpretable in the course of everyday activities. Contextual aspects of interactional exchanges make any of it unique and requiring different interpretations and actions. A robotic socially believable system must be able to discriminate among the infinities of contextual instances and assign to each their unique meaning. This book reports on the last research efforts in making "natural" human interactional exchanges with social robotic autonomous systems devoted to improve the quality of life of their end-users while assisting them on several needs, ranging from educational settings, health care assistance, communicative disorders, and any disorder impairing either their physical, cognitive, or social functional activities.

Keywords Social robotic autonomous systems · (social, physical and organizational) context · End-users' quality of life · Interactional exchanges

2.1 Introduction

The human ability to merge information from different sensory systems offers more accurate and faster competencies to operate in response to the environmental stimuli [1]. The integration of different signals in an unique percept is especially

A. Esposito (✉)
Dipartimento di Psicologia and IIASS, Seconda Università di Napoli,
Caserta, Italy
e-mail: iiass.annaesp@tin.it

L.C. Jain
Faculty of Education, Science, Technology & Mathematics,
University of Canberra, Canberra, Australia
e-mail: jainlc2002@yahoo.co.uk

© Springer International Publishing Switzerland 2016
A. Esposito and L.C. Jain (eds.), *Toward Robotic Socially Believable Behaving Systems - Volume II*, Intelligent Systems Reference Library 106,
DOI 10.1007/978-3-319-31053-4_2

appreciated in noisy environments with corrupted and degraded signals [20]. Research in neuroscience had proved that audiovisual, visual-tactile and audio somatic sensory inputs are constantly synchronized and combined into a reasoned percept [3, 33]. For example, in speech, the influence of visual on auditory signals perception is proved by the McGurk effect [28]. Recent investigations on the emotional labelling of only audio, mute video, and combined audio/video stimuli proved that the human processing of emotional information is strongly affected by the context to the extent that, depending on the participant' language, culture, and her knowledge of a given foreign language, her performance on an emotion recognition task, accuracy may become significantly worse when the visual signal (emotional facial expressions) is added to audio (emotional vocal expressions, [9, 11–13, 16].

Facial expressions, head, body and arms movements (grouped under the name of gestures) all potentially provide information to the communicative act, supporting the interactional exchange and allowing interactants to add a rich variety of contextual information to their messages including (but not limited to) their psychological state, attitude. Gestures have been shown to vary in size, redundancy, and complexity, depending on the grounding status of the information encoded and/or the meaning attributed to the message. Gestures act in partnership with speech, building up shared knowledge and meanings when the interactional exchange is successful [10, 15, 23].

Psycholinguistic studies have confirmed the partnership nature of verbal and non-verbal signals in human interaction demonstrating that the understanding of a message results from the integration of multi-sensory features appropriately distributed along the interaction [30].

In addition, it has been suggested that interactive communication is emotionally driven [8] and that the encoding and decoding procedures exploited by humans to express and/or read emotions are fundamental to secure the social quality and the cognitive functioning of a successful interactional exchange.

Another crucial aspect of multimodal communication is the relationship between paralinguistic and extra-linguistic information (such as speech pauses, head nodding). Psycholinguistic studies have shown that there exists a set of non-lexical expressions carrying specific communicative values (expressing for example turn-taking and feedback mechanism regulations) such as empty and filled pauses and other hesitation phenomena [14, 15, 17]. It has been shown that pauses (holds) in gestures plays similar communicative values and synchronize with paralinguistic information [10].

It can be concluded that the verbal and nonverbal communication modes jointly cooperate in assigning semantic and pragmatic contents to the conveyed message by unraveling the participants' cognitive and emotional states and exploiting this information to tailor the interactional process. These modes exploit multimodal social signals and are tailored to the contextual instance in which they are explicated.

The huge demand for complex autonomous systems able to assist people on several needs had produced a consistent number of EU and overseas funded projects such as (a) ERICA (www.jst.go.jp/erato/ishiguro/en/) a conversational android with human appearance aiming to interact with humans through multimodal social signals such as face, speech, and body movements; (b) MUMMER a humanoid robot (based on Aldebaran's Pepper platform, http://www.dcs.gla.ac.uk/vincia/?page_id= 116) engaging people in dynamic environments, such as shopping malls; (c) TESLA—An Adaptive Trust-based e-assessment System for Learning (www. open.ac.uk/iet/main/research-innovation/research-projects/adaptive-trust-based-e-assessment-system-learning-tesla); (d) ALIZ-E (http://www.aliz-e.org/) engaging diabetic children in a series of real-world situations (partially) WOZ-simulated implemented using the NAO robot (http://www.aldebaran-robotics.com/en) [25, 26]; and several more, all exemplifying the huge research efforts in implementing socially believable assistive technologies. However, these projects have been characterized by a scarce care to what effectively would have been end-users' requirements and expectations to qualify as "socially behaving" agents providing "*social/physical/psychological/assistive ICT services. In particular, the term "social robotics" envisions a "natural" interaction of such devices with humans, where "natural" is interpreted as the ability of such agents to enter the social and communicative space ordinarily occupied by living creatures*" [7, p. 2]. In addition really few has been made to account for "*how the interaction between the sensory-motor systems and the inhabited environment (that includes people as well as objects) dynamically affects/enhances human reactions/actions, social perception and meaning-making practices*" [6, p. 6].

To account for the above mentioned problems it is needed an all-embracing prospect pushing the designer to contemplate the system's behavior and appearance (in order for the system to be social), the trustworthiness the user put into it (in order to be emphatic) taking into account the contextual instance (the scenario) and the system' functionalities required in each situation, as well as, the individual's social rules and cognitive competencies [5].

In the field of Human Robot Interaction, such an approach, will require investigations on the cognitive architectures and cognitive integrations needed for accounting of human behavior across different domains, and inherently of the behavior humans engage with a system that, as much as complex and autonomous can be, can offer only a sub-optimal interaction process (see key activities of the topic group Natural Interaction with Social Robots http://homepages.stca.herts.ac.uk/~comqkd/TG-NaturalInteractionWithSocialRobots.html, [6, 7, 19]).

To date there have been relatively few efforts assessing human interactional exchanges in context in order to develop complex autonomous systems able to detect user's trust and mood, rise emphatic feeling, and take actions to provide help. In addition there are no standards for the development of more 'satisfying' complex autonomous systems that account for user's expectations and requirements in a structured manner. Although there have been efforts in providing suggestions for potential solutions [2, 18, 32] this issue is at a research stage [7, 19].

Generally the development and assessment of complex autonomous systems is tackled using two different approaches: user's self reports and performance based measures. The user's self reports can be highly criticized because of the user's difficulty to accurately describe her/his expectations and requirements, being them generally technologically naïve and suspicious. Performance-based measures can be considered more reliable since they require the execution of specific tasks assessed by a trained evaluator. Nevertheless, these tasks are generally carried out under artificial conditions and require extensive equipment, well defined environmental context and time consuming evaluation procedures that do not value the daily spontaneous activity producing biases in the collected measures. Despite its importance, generally these systems are unable to being context-aware, to adapt to user's preferences and very distinct needs and to correctly interpret all user's actions.

The research papers proposed in this book investigate the features that are at the core of human interactions and provide attempts to model the cognitive and emotional processes involved in order to design and develop complex autonomous system prototypes able to simulate the human's ability to decode and encode social cues while interacting.

2.2   Content of the Book

The research objectives proposed in this book can be interpreted as a meta-methodology aimed to investigate features that are at the core of human interactional exchanges and model the cognitive and emotional processes involved in interactions, in order to design and develop complex autonomous system prototypes able to socially behave in (at the least) specific scenarios. The attention is focused on the analysis and modeling of social behavioral features and human ability to decode and encode social cues while interacting. Behavioral data (speaking, body movements, facial, vocal and gestural emotional expressions) are gathered from healthy and communicative or socially impaired participants. This require the definition of behavioral tasks that serve both to detect changes in the healthy, as well as, impaired perception of social cues. Specific scenarios are proposed for these tasks intended to assess the users' attitude, acceptance, and trustworthiness toward a robotic system considering its emphatic and social competencies, as well as appearance. The collected data are used to gain knowledge on how behavioral and interactional features are affected by individual characteristics and personalities, contextual instances, and environmental perceptual features. Hopefully, these investigations will guide on which human-like social characteristics and appearance (physically embodied or virtual intelligent agent?) a complex autonomous system should exhibits to gain the users' trust and acceptance as a socially behaving agent.

To this aim, the book includes nine investigations on the mathematical modeling of social signals and context embedded in interactional exchanges. The second chapter by Maldonato and Dell'Orco [27] affords one of the most debated issue in artificial intelligence: "*the possibility of reproducing in an artificial agent (based on formal algorithms) some typically human capacities (based on natural logic algorithms) such as consciousness, the ability to deliberate and make moral judgments*" [p. 1]. The authors are very desecrators to the point of asserting that "*clarifying consciousness mechanisms of artificial organisms could help us to discover what we still ignore about neurobiological consciousness*" [p. 2]. This extreme in the quest for equipping machines with human level automaton intelligence, consciousness, and intuition, leave us with the question on whether we really want conscious and intuitive artificial agents. The answer is given by the contribution of Gnjatović and Borovac [21] which propose the implementation of conscious-like conversational agents, implicitly answering to the question with a discussion on which are the limitation for implementing consciousness features in mathematical prototypes. These investigations clearly suggest that the aims of the research on social robotics is to implement natural interactions with such social agents. The contribution of Vogel [35] covers one aspect of this sociability proposing an investigation aimed to the *"understanding of natural dialogue"* in order to *"fully inform the construction of believable artificial systems that are intended to engage in dialogue with a manner close to human interaction in dialogue"* [p. 1]. Harrington et al. [22] follow a similar vision considering *"the relevance of context and experience for the operation of historical sound changes"* [p. 1]. The contribution of Clavel et al. [4] is an original survey on which competencies an artificial agent must exploit to maintain, when interacting, users' engagement. The focus is on both users' attentional and emotional involvement. op den Akker's et al. [31] contribution propose *"Kristina, a personal digital coaching system built to support and motivate users to live a balanced and healthy lifestyle"* [p. 14]. The authors are aware of dangers and objections that can be raised by modeling interactional persuasive features and provide a very interesting discussion on these aspects. The contribution of Vinciarelli [34] discuss on how *"endowing machines with social perception"*, in particular *"by providing a simple conceptual model of social perception and by showing a few examples related to Automatic Personality Perception, the task of predicting how people perceive the personality of others"* [p. 1]. Finally, the last two contributions surpass dyadic interactional features considering to model either multimodal and multiparty interactions in educational settings, as in the work of Koutsombogera et al. [24], or to detect abnormal behavioral patterns in crowd scenarios, as in the contribution of Mousavi et al. [29].

2.3 Conclusions

The readers of this book will get a taste of the major research areas in modeling social signals and contextual instances of interactional exchanges in different scenarios for implementing robotic socially believable behaving systems. This research should result in a series of theoretical and practical advances in the field of cognitive, and social psychology such as: (1) Repertories of social signals better illustrating the cognitive, semantic, emotional and semiotic mechanisms essential for successful interactional human-machine exchanges; (2) Models for representing data, reasoning, learning, planning, and decision making, as well as, individual/group behavior analysis models in multilingual and cross-cultural contexts; (3) The identification of new interactional persuasive and affective strategies and contextual instances calling for their use. Considering technological issues, the present research must lead to (1) New computational approaches and departures from existing cognitive frameworks and existing algorithmic solutions such as dynamic Bayesian networks, long short-term memory networks, and fuzzy models of computation; (2) The implementation of behaving ICT systems of public utility and profitable for a living technology that simplifies user access to future, remote and nearby social services encompassing language barriers and cultural specificity; (3) Market applications such as: context-aware avatars replacing human in high risk tasks, companion agents for elderly and impaired people, socially believable robots interacting with humans in extreme, stressful time-critical conditions, future smart environments, ambient assistive living technologies, computational intelligence in games/storytelling, embodied conversational avatars, and automatic healthcare and education services.

References

1. Block N (1995) The mind as the software of the brain. In: Smith EE, Osherson DN (eds) Thinking. MIT Press, Cambridge, pp 377–425
2. Brandão Moniz A (2010) Anthropocentric-based robotic and autonomous systems: assessment for new organisational options. IET Working Papers Series No. WPS07/2010
3. Callan DE et al (2003) Neural processes underlying perceptual enhancement by visual speech gestures. NeuroReport 14:2213–2218
4. Clavel C, Cafaro A, Campano S, Pelachaud C (2016) Fostering user engagement in face-to-face human-agent interactions: a survey. This volume
5. Davis MH (1983) Measuring individual differences in empathy: evidence for a multidisciplinary approach. J Personal Soc Psychol 44:113–126
6. Esposito A, Esposito AM, Vogel C (2015) Needs and challenges in human computer interaction for processing social emotional information. Pattern Recognit Lett 66:41–51
7. Esposito A, Fortunati L, Lugano G (2014) Modeling emotion, behaviour and context in socially believable robots and ICT interfaces. Cognit Comput 6(4):623–627
8. Esposito A (2013) The situated multimodal facets of human communication. In: Rojc M, Campbell N (eds) Coverbal synchrony in human-machine interaction, ch. 7. CRC Press, Taylor & Francis Group, Boca Raton, pp 173–202
9. Esposito A, Esposito AM (2012) On the recognition of emotional vocal expressions: motivations for an holistic approach. Cognit Process 13(2):541–550
10. Esposito A, Esposito AM (2011) On speech and gesture synchrony. In: Esposito A et al (eds) Communication and enactment. LNCS, vol 6800. Springer, New York, pp 252–272
11. Esposito A, Riviello MT (2011) The cross-modal and cross-cultural processing of affective information. In: Apolloni B et al (eds) Frontiers in artificial intelligence and applications, vol 226. IOSpress, Amsterdam, pp 301–301
12. Esposito A (2009) The perceptual and cognitive role of visual and auditory channels in conveying emotional information. Cognit Comput J 1(2):268–278
13. Esposito A, Riviello MT, Bourbakis N (2009) Cultural specific effects on the recognition of basic emotions: a study on Italian Subjects. In: Holzinger A, Miesenberger K (eds) USAB 2009. LNCS, vol. 5889. Springer, Berlin, pp 135–148
14. Esposito A (2008) Affect in multimodal information. In: Tao J, Tan T (eds) Affective information processing. Springer, Heidelberg, pp 211–234
15. Esposito A, Marinaro M (2007) What pauses can tell us about speech and gesture partnership. In: Esposito A et al (eds) Fundamentals of verbal and nonverbal communication and the biometric issue. NATO series human and societal dynamics, vol 18. IOS press, The Netherlands, pp 45–57

16. Esposito A (2007) The amount of information on emotional states conveyed by the verbal and nonverbal channels: some perceptual data. In: Stilianou Y et al (eds) Progress in nonlinear speech processing. LNCS, vol 4391. Springer, Heidelberg, pp 245–268
17. Esposito A (2006) Children's organization of discourse structure through pausing means. In: Faundez_Zanuy M et al (eds) Nonlinear analyses and algorithms for speech processing. LNCS, vol 3817. Springer, New York, pp 108–115
18. Feil-Seifer D, Skinner K, Matarić MJ (2007) Benchmarks for evaluating socially assistive robotics. Interact Stud: Psychol Benchmarks Hum-Robot Intreact 8(3):423–429
19. Fortunati L, Esposito A, Lugano G (2015) Beyond industrial robotics: social robots entering public and domestic spheres. Inf Soc: Int J 31(3):229–236
20. Garrigan P, Kellman PJ (2008) Perceptual learning depends on perceptual constancy. PNAS 105(6):2248–2253
21. Gnjatović M, Borovac B (2016) Toward conscious-like conversational agents. This volume
22. Harrington J, Kleber F, Reubold U, Stevens M (2016) The relevance of context and experience for the operation of historical sound change. This volume
23. Kendon AG (2005) Visible action as utterance. University Cambridge Press, New York
24. Koutsombogera M, Deligiannis M, Giagkou M, Papageorgiou H (2016) Towards modelling multimodal and multiparty interaction in educational settings. This volume
25. Kruijff-Korbayová I, Athanasopoulos G, Beck A, Cosi P, Cuayáhuitl H, Dekens T, Enescu V, Hiolle A, Kiefer B, Sahli H, Schröder M, Sommavilla G, Tesser F, Verhelst W (2011) An event-based conversational system for theNAO robot. In: IWSDS 2011. Granada, Spain
26. Kruijff-Korbayová I, Cuayáhuitl H, Kiefer B, Schröder M, Cosi P, Paci G, Sommavilla G, Tesser F, Sahli H, Athanasopoulos G, Wang W, Enescu V, Verhelst W (2012) Spoken language processing in a conversational system for child-robot interaction. In: Workshop on child-computer interaction
27. Maldonato M, Dell'Orco S (2016) Adaptive and evolutive algorithms: a natural logic for artificial mind. This volume
28. McGurk H, MacDonald J (1976) Hearing lips and seeing voices. Nature 264:746–748
29. Mousavi H, Galoogahi HK, Perina A, Murino V (2016) Detecting abnormal behavioral patterns in crowd scenarios. This volume
30. Munhall KG, Jones JA, Callan DE, Kuratate T, Vatikiotis-Bateson E (2004) Visual prosody and speech intelligibility. Psychol Sci 15(2):133–137
31. op den Akker HJA, Klaassen R, Nijholt A (2016) Virtual coaches for healthy lifestyle. This volume
32. Petrelli D, Not E (2005) User-centred design of flexible hypermedia for a mobile guide: reflections on the hyperaudio experience. User Model User-Adapt Inter 15(3–4):303
33. Stekelenburg JJ, Vroomen J (2007) Neural correlates of multisensory integration of ecologically valid audiovisual events. J Cognit Neurosci 19(12):1964–1973
34. Vinciarelli A (2016) Social perception in machines: The case of personality and the Big-Five traits. This volume
35. Vogel C (2016) Communicative sequences and survival analysis. This volume

Chapter 3
Adaptive and Evolutive Algorithms: A Natural Logic for Artificial Mind

Mauro Maldonato and Silvia Dell'Orco

Abstract This paper focuses on one of the biggest challenges we face: the possibility of reproducing in an artificial agent (based on formal algorithms) some typically human capacities (based on natural logic algorithms) such as consciousness, the ability to deliberate and make moral judgments. Recent evidences arising from dynamic systems theory and statistical learning, from the psychobiology of development and molecular neuroscience are overcoming some of the fundamental assumptions of artificial intelligence and the cognitive science of the last 50 years. From the molecular level to the social one, these new approaches analyze and exploit the structure of complex causal systems physically incorporated and integrated with the environment, setting the stage for the emergence of organisms capable of adaptive flexibility and intelligent behavior.

Keywords Artificial mind · Consciousness · Decision-making · Natural logic · Algorithms · Intuition

3.1 Introduction

Are increasing those who believe that the encounter between man and powerful computers will generate, in the near future, organisms capable of going beyond the simulation of brain function: hybrids that will learn from their internal states, interpret the data of reality, establish their own goals, converse with humans, decide on the basis of its own 'value system' [1–3]. Sooner than we imagine, these organisms could acquire spheres of autonomy increasingly large through self-conservative

M. Maldonato (✉)
Department of European and Mediterranean Cultures (DICEM),
University of Basilicata, Matera, Italy
e-mail: mauro.maldonato@unibas.it

S. Dell'Orco
Department of Human Sciences (DISU), University of Basilicata, Potenza, Italy
e-mail: silvia.dellorco@unibas.it

© Springer International Publishing Switzerland 2016
A. Esposito and L.C. Jain (eds.), *Toward Robotic Socially Believable Behaving Systems - Volume II*, Intelligent Systems Reference Library 106,
DOI 10.1007/978-3-319-31053-4_3

instances, hierarchies of values, perhaps an ethic based on the 'freedom'. In other words, developments in artificial intelligence and in artificial life may take us out of the symbolic-formals domains to achieve a brain with central control functions and higher cognitive properties and, therefore, an intelligence that processes symbolic and abstract functions as those of the biological brain. From that point to spontaneity, creativity and consequently the ability to make choices is short.

It will be very difficult to replicate the huge work of evolution and still define as 'experience' an information processing (albeit sophisticated), or still call emotions' experiences as pleasure or pain. In fact, the experiences of consciousness must be learned directly; they don't depend on relational aspects and, finally, they are intimate and private. But can we really exclude that one day these entities will be capable of taking decisions, initiatives and have discernment? We can't. At least in theory, this could just happen with thinking machines. Rather, precisely the construction of artifacts, although still equipped with a primitive awareness, could help us to formulate the right questions to understand the mechanisms of consciousness and other mental functions that still elude us. Clarifying such structures and dynamics, may also light the shadows that obscure our knowledge.

3.2   Can a Machine Become Self-Aware?

There is no doubt that the artificial consciousness is not yet considered, at least outside of science fiction, a discipline in its own right. Nevertheless, despite its elusiveness, the theme of consciousness has been reinterpreted in the light of the progress of Artificial Intelligence, on the one hand, and of neurobiology and cognitive sciences on the other [4, 5].

Paradoxically, the problem of consciousness is becoming more urgent with the progress of our possibility to investigate the brain [4]. There is no reason a priori to deny that an artificial creature, able to reproduce the basic mechanisms of the human brain, may correspond to a conscious subject.

The understanding of the biological mechanisms underlying the emergence of consciousness is not clear at all. In any case, within the themes of artificial consciousness, we should ask ourselves what we mean by consciousness.

As a neurobiological phenomenon distinct from awareness, consciousness originates in the cortical-subcortical space, even if it is only in the cerebral cortex that the experience of time is realized, that is, the unmistakable individual impression of continuous past experiences that is bound together with future expectations [6, 7].

It seems to be a general consensus that at the basis of consciousness there is synchronization between different cerebral regions [8]. However, the question remains open as to the nature of the passage from the neuronal level to that of perception and, finally, consciousness. Varela [9] has long insisted on the necessity of considering consciousness as an emerging phenomenon, in which local events can give rise to properties or global objects in a reciprocal causal co-involvement. More recent theories on consciousness hypothesize a minimum necessary amount of time for the emergence of neural events that connect themselves to a cognitive event [10]. This temporality can plausibly be attributed to long-range cerebral integration linked to diffuse synchrony: an event that would shed light on phenomenological invariants, restoring tangible experiential content to the synchronization process: i.e. that mechanism of temporal unification of neuronal activity that would synchronize pulses in average fluctuations of 40 Hz, which would unify part of the existing information in a coherent sense [11]. The synchronization of the discharges would entrust the psychic content to the working memory, while consciousness would reply in an intermediate zone of representations, displacing it between a lower sensory level and a higher level of thought [12].

For a long time, scholars focused on the concept of the unitarity and the permanence of consciousness in time [13]. Today, instead, numerous studies show that consciousness is a plural process that encompasses different contents in itself simultaneously, each element of which has its own intentionality [14].

But what are the biophysical mechanisms of the unified experience of consciousness? And how does this internal plurality unify the different contents? There seem to be two possible models. The first model hypothesizes that consciousness is generated by a central neural system, in which duly integrated information is first represented and then brought to consciousness [15]. In this schema consciousness appears to be the result of the work of the central neural system that generates different contents and representations, a phenomenon taking place exclusively in the brain. In the second model [16] the simultaneous co-activation of the contents—specific and aware perceptions or engrams of memory that are part of the stream of consciousness (like the sight of a "blood orange", "the smell of a rose" or "a feeling sad")—generated by distributed structures in the brain are believed to give rise, ultimately, to the phenomenon of consciousness. Consciousness would in this way be generated by distributed cerebral mechanisms—both cortical and subcortical—the contents of which, each element being independent one from the other, are exposed to intrasensory and intersensory (environmental) influences [17]. The contents of the distributed cerebral mechanisms and the intrasensory and intersensory influences affect each other reciprocally and thus co-determine conscious experiences. It is in this fine line that the distinction between a unitary model and a plural model of consciousness lies. According to the unitary model, the conscious experience is a process of creation in which the brain acts simultaneously treating the different information; therefore, this is realized exclusively in the brain. On the contrary, according to the plural model of consciousness—consciousness would be the effect of a series of phenomenal emerging elements, each one of them generated by a precise brain mechanism—the contents are independent of each other [18, 19] and give rise to a asynchronous and multiple system of micro-science, in which consciousness is not a faculty of synthesis hierarchically superior, but a myriad of integration of quantitative and qualitative phenomena.

Ramachandran [20] has a number of times discussed the plausibility of a model of consciousness that integrates visual, auditory and tactile perceptions with proprioceptive, exteroceptive and interoceptive experiences. These individual spheres, in a relatively independent way, can be altered or neutralized without influencing

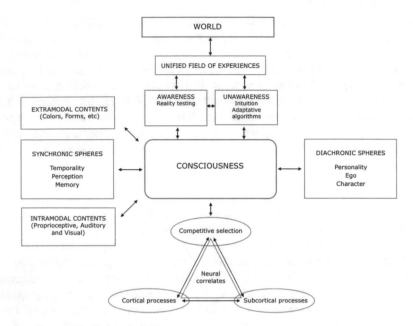

Fig. 3.1 We propose here a unified spectrum model: a plural embodied perspective of consciousness

the other spheres. The modularity of functions, though exposed to intra-sensory and inter-sensory influences, put away the contents aware from constraints and interdictions. Studies [21] on the deficits caused by brain's lesions on the level and kind of functional specialization, such as language or spatial cognition, and cerebral localization have shown that the brain works on a large scale, between procedures and domains that are reflected in specific anatomical districts (primary visual processing in the occipital cortex, auditory processing in the temporal cortex, planning and memory processing in the frontal cortex), while specific functions are realized in well-demarcated anatomical districts and locations (for example, the visual motor function takes place in area V5 and that of color in V4). The zones of the brain that program particular informational content are those in which the contents come into consciousness (Fig. 3.1).

Zeki and Bartels [18] showed that several events of a visual scene, presented simultaneously, are not perceived with the same duration. Hence they inferred the existence of a system asynchronous multiplex, which seems to prove that consciousness is the integrated result of countless micro events and not a unitary faculty [14].

But how can these multiple neural events restore to us the impression of a unitary subjectivity? And which paths lead to the composition of the Self and of consciousness? Concepts such as "unitary subjectivity" and "Self"—i.e. the ability to experience phenomena as perceptions, thoughts or emotions—remain problematic. Indeed, both concepts appear abstract, i.e. without precise correspondences in the functions of the mind and, more particularly, of correlations with the structures of the central nervous system. Here, we will limit ourselves to affirming that the Self emerges when individual events produced by the brain are sufficiently representational, coherent and close-knit, that is, capable to reach a discrete threshold of sensible mental contents. In the absence of neurological and psychiatric disorders, we experience a structured world of distinct objects ordered in space, organized according to regularities and contents within meaningful spatial-temporal schemas: extramodal contents (colors, forms, etc.) and intramodal contents (proprioceptive, auditory and visual) [22]. In reality representational cohesion is not an invariant characteristic of conscious experience, but the result of a selection through which the brain searches for the path of its own integration [23]. Ultimately, the Self has to do with a regulatory activity of consciousness that processes and maintains the distributed cortical and sub-cortical activity, in an interweaving of local contents in communication with each other. In this schema, consciousness appears not as a hierarchical entity, but as a multiple horizontal entity, whose representational cohesion is carried out by thalamic-cortical and cortico-cortical circuits [24]. All conscious experiences, beginning with those that are qualitative (qualia), become unified within the field of consciousness. In this sense, if our consciousness is determined by the play between these innumerable dynamics, the unitarity of consciousness follows the subjectivity, because there is no qualitative subjectivity without unity.

The issue of conscious subjectivity goes beyond the search for its neuronal correlates and even beyond the conceptual contraposition between consciousness and the unconscious. In this sense, an in-depth study of consciousness requires multi-level explanatory criteria: a quantitative-categorical criterion (attention, alertness, sleep, and coma); a qualitative-dimensional criterion (subjective experiences such as sensations, thoughts, and emotions); and a final criterion for the analysis of the different synchronic (the field of consciousness) and diachronic (the I and personality) types and levels of consciousness, as the clouding of consciousness, confusional state, obtundation, stupor and so on [19].

We do not know whether there are limits to this ability to be conscious or whether it can have a further expansion: in other words, the activation of neuronal spontaneous mechanisms, a sort of "neuronal avalanche", invests the brain through waves of electricity that form complex structures and extend the boundaries of our mental life [25]. It is necessary to emphasize that being conscious and being alive are different things. To be alive means to be made up of molecules based on DNA replication, be conscious means to be able to experience the world [26].

Based on the current knowledge [27, 28], nothing binds the DNA structure or the carbon atom one to our conscious state. And if this is true it would mean that you could assume and imagine the way in which an artificial structure can produce a subject with conscious experiences, in the sense that clarifying consciousness mechanisms of the artificial organisms could help us to discover what we still ignore about neurobiological consciousness.

So far, the design of artifacts did not require the understanding of consciousness. But, nowadays the robots begin to approach the human being—in the sense of their computing capacity and physical structure—and that means that the problem cannot be avoided. For example, an embryo, before the appearance of the nervous system, can't be considered a conscious subject; but after a sufficient period of time, the human being will result conscious [29].

The mind is no longer an empty box that receives sounds and images from the outside world, but it is a portion of the external world that found in itself its own unity.

Actually is necessary to revisit—and update—our philosophical categories, that may suggest to engineers the methodological and technical processes to achieve one of the most attractive targets proposed by science in the XXI century: understanding who we are through the construction of a machine that can tell what happens inside the human being and how he has feelings.

The final act of the great drama ($\delta\varrho\hat{\alpha}\mu\alpha$) of the human being, who wants to understand himself, cannot be written by the human sciences but through artificial intelligence that would allow us, through artificial consciousness, to be able to communicate with the subjects constructed by us.

3.3   Decision Making and Intuition Algorithms

Strictly linked to the issue of artificial consciousness is the problem of the artificial decision making. Decision-making is a significant research area of artificial intelligence. Making decisions is the main characteristic of any virtual character. The latter generally has a limited number of actions it can choose to perform. There are many different decision-making algorithms [30] for these virtual characters, some based on planning, others on reactivity, and others again on a their combination. In most part of situations in which virtual characters have to make a decision, the decision algorithms must be quick enough to allow real-time choices. Here, an important challenge is the capacity to join reactivity and planning in order to be able to act without hesitations. The greater limitation to reactivity is that actually the virtual character cannot anticipate the consequences of a particular behavior. They aren't provided of a meta-critical reasoning. In fact, decision-making algorithms not include intuitions, feelings, emotions and spheres of free will that lead to quick and adaptive decisions [31]. Actually, we only know that intuition is a form of instinctive and unconscious knowledge that enables us—not based on logical deductive processes—to face up the unexpected events in a new and often resolutive way for survival. Intuition comes into play in situations in which temporal and cognitive-computational constraints prevent us from reflecting on or evaluating all the data at our disposal [32]. Since the dawn of time, being able to decipher rapidly the intentions of whoever you are confronting increases your chances of survival. This explains why often the first instants of an encounter can reveal more than hours of conversation [33, 34]. Besides, in the various cultures the world over, the ability to interpret non-verbal signals is of enormous importance [35]. In reality, the vast majority of human decisions are intuitive and unconscious; requiring only limited mental input [36]. These enable us to elaborate great quantity of information rapidly and without great efforts. In any situation in which we have achieved a high degree of experience, an incalculable stock of information has been accumulated at the level of "gut instinct". Over the last twenty years there has been a remarkable surge in the volume of research [37] on the instinctive mechanisms, which show that from our deep memory emerge response schemes to reality issues, on the basis of analogies with past experiences. This process can reveal unexpected solutions for the problems that we will have to face, that may be useful to achieve a rigorous definition of the term 'intuition'.

What is an intuition about? Is it creativity, tacit knowledge, implicit learning and memory, sixth sense, heuristics, emotional intelligence? It's really difficult to say. Intuition has common characteristics with all the above, but also with other definitions. The word 'insight', for example, often considered a synonymous of 'intuition', indicates the sudden comprehension of a problem or a problem-solving strategy, the 'Eureka!' moment that arrives after a period of more or less conscious incubation, unblocking the solution to the problem [38]. In truth an intuition almost always has its somatic correlation in a sensation in the pit of the stomach, coming all of a sudden. But what lies behind all this? One stimulating hypothesis (not without metaphysical overtones) is that a crowd of "cognitive workers" are engaged every day in the subterranean regions of our mind, far from the light of consciousness, elaborating extraordinary quantities of information that involve the implicit memory, heuristics, spontaneous inferences, emotions, creativity, and much else [39]. The complexity and the difficulties within the AI research program was already clear to von Neumann [40] who saw clearly the obstacles inherent to a cybernetic model of human behavior. The question before us, however, is to overcome the elementary functions, mechanically predetermined, that exhibit intuitive behavior patterns. Of course, we are forced to deal with a difficult issue that we humans have always solved by evolution: being able to distinguish what is important from what is irrelevant, sharpen our discriminative capacity, keep us open to other possible explanations, and more.

3.4 Conclusions

At the beginning of the twentieth century, psychoanalysis has brought to light the conflict between unconscious forces and rationality [41]. It showed us how freedom, decisions and moral choices are the fatiguing expression of this conflict. Today, more than a century later, the neurosciences [16] try to clarify the nature of these matters, on the base of experimental observations. At the height of these challenges, science will face the hardest problem, namely that of artificial agents able to deliberate and make moral judgments and act on the basis of their values [42].

Intuition takes place almost instantaneously and is made up of a set of emotional and somatic processes, without any role being played (at least apparently) by rational, conscious thought [43]. How these human processes can be processed by silicon structures is the great challenge. One of the main obstacles is that non-logical reasoning drives these processes, while most of the AI models are governed by logic.

References

1. Kurzweil R (1990) The Age of Intelligent Machines. MIT Press, Cambridge
2. Shen S (2011) The curious case of human-robot morality. In: Proceedings of the 6th international conference on human robot interaction, Lausanne, Switzerland, Mar 6–9
3. Powers TM (2011) Incremental machine ethics. IEEE Robot Autom Mag 18:51–58
4. Dehaene S (2014) Consciousness and the brain: deciphering how the brain codes our thoughts. Viking Penguin, New York
5. Batterink L, Neville HJ (2013) The human brain processes syntax in the absence of conscious awareness. J Neurosci 33:8528–8533
6. Tononi G (2004) An information integration theory of consciousness. BMC Neurosci 5:42
7. Dehaene S, Charles L, Remi King J, Marti S (2014) Toward a computational theory of conscious processing. Curr Opin Neurobiol 25:76–84
8. Sergent C, Baillet S, Dehaene S (2005) Timing of the brain events underlying access to consciousness during the attentional blink. Nat Neurosci 8:1391–1400
9. Varela FJ (1996) Neurophenomenology: a methodological remedy for the hard problem. J Conscious Stud 3(4):330–350
10. Dennett D, Kinsbourne M (1992) Time and the observer: the where and the when of the consciousness in the brain. Behav Brain Sci 15:183–247
11. Crick F, Koch C (2003) Framework for consciousness. Nat Neurosci 6(2):119–126
12. Dehaene S (2008) Conscious and nonconscious processes. In: Engel C, Singer W (eds) Distinct forms of evidence accumulation better than conscious? decision making, the human mind, and implications for institutions. MIT Press, Cambridge, pp 21–50
13. McGinn C (1991) The problem of consciousness. Blackwell, Oxford
14. Zeki S (2003) The disunity of consciousness. Trends Cogn Neurosci 7:214–218
15. Baars B (1997) In the theater of consciousness: the workspace of the mind. Oxford University Press, NY
16. Gazzaniga MS (2004) The new cognitive neurosciences. MIT Press, Cambridge
17. Menon S (2014) Brain, self and consciousness: explaining the conspiracy of experience. Springer, New Delhi
18. Zeki S, Bartels B (1998) The asynchrony of consciousness. Proc R Soc B 265:1583–1585
19. Maldonato M (2015) The archipelago of consciousness. The invisible sovereignty of life. Sussex Academic Press, Brighton
20. Ramachandran V (2004) The emerging mind. Profile Books, London
21. Geschwind N, Galaburda AM (1987) Cerebral lateralization: biological mechanisms, associations and pathology. MIT Press, Cambridge
22. Dehaene S, Kerszberg M, Changeux JP (1998) A neuronal model of a global workspace in effortful cognitive tasks. Proc Natl Acad Sci USA 95:14529–14534
23. Sackur J, Dehaene S (2009) The cognitive architecture for chaining of two mental operations. Cognition 111:187–211
24. Dehaene S, Changeux JP (2011) Experimental and theoretical approaches to conscious processing. Neuron 70:200–227
25. Shew WL, Yang H, Petermann T, Roy R, Plenz D (2009) Neuronal avalanches imply maximum dynamic range in cortical networks at criticality. J Neurosci: Off J Soc Neurosci 29:15595–15600
26. Molyneux B (2012) How the problem of consciousness could emerge in robots. Minds Mach 22:277–297
27. Kurzweil R (2012) How to create a mind: the secret of human thought revealed. Viking Books, New York
28. Kaku M (2008) Physics of the impossible. Doubleday, New York
29. Gamez D (2012) Empirically grounded claims about consciousness in computers. Int J Mach Conscious 4:421–438
30. Nisan N, Roughgarden T, Tardos E, Vazirani VV (2007) Algorithmic game theory. Cambridge University Press, New York

31. Adiandari AM (2014) Intuitive decision making: the intuition concept in decision making process. Int J Bus Behav Sci 4(7):1–11
32. Maldonato M, Dell'Orco S (2011) Natural logic., Exploring decision and intuitionSussex Academic Press, Brighton
33. Brown JW, Braver TS (2005) Learned predictions of error likelihood in the anterior cingulate cortex. Science 307:1118–21
34. Maldonato M, Dell'Orco S (2012) The predictive brain. World Futur Routledge 68:381–389
35. Tsiamyrtzis P, Dowdall J, Shastri D, Pavlidis IT, Frank MG, Ekman P (2007) Imaging facial physiology for the detection of deceit. Int J Comput Vis 71(2):197–214
36. Gigerenzer G (2007) Gut feelings: the intelligence of the unconscious. Viking, New York
37. Kahneman D (2011) Thinking. Fast and Slow, Farrar, Straus and Giroux, New York
38. Oliverio A, Maldonato M (2014) The creative brain. In: CogInfoCom, 5th IEEE international conference on cognitive info-communications, Novemb 5–7: 527–532
39. Baars BJ, Gage NM (2007) Cognition, brain & consciousness: an introduction to cognitive neuroscience. Academic Press (Elsevier), San Diego, Calif
40. von Neumann J (1958) The computer and the brain. Yale University Press, New Haven
41. Epstein S (1994) Integration of the cognitive and the psychodynamic unconscious. Am Psychol 49(8):709–724
42. Wallach W, Franklin S, Allen A (2010) A conceptual and computational model of moral decision making. Hum Artif Agents 2(3):454–485
43. Newell BR, Shanks DR (2014) Unconscious influences on decision making: a critical review. Behav Brain Sci 37(1):1–61

Chapter 4
Toward Conscious-Like Conversational Agents

Milan Gnjatović and Branislav Borovac

Abstract Although considerable effort has been already devoted to studying various aspects of human-machine interaction, we are still a long way from developing socially believable conversational agents. This paper identifies some of the main causes of the current state in the field: (i) socially believable behaviour of a technical system is misinterpreted as a functional requirement, rather than a qualitative, (ii) the currently prevalent statistical approaches cannot address research problems of managing human-machine interaction that require some sort of contextual analysis, and (iii) the structure of human-machine interaction is unjustifiably reduced to a task structure. In addition, we propose a way to address these pitfalls. We consider the capability of a technical system to simulate fundamental features of human consciousness as one of the key desiderata to perform socially believable behaviour. In line with this, the paper discusses the possibilities for the computational realization of (iv) unified interpretation, (v) learning through interaction, and (vi) context-dependent perception in the context of human-machine interaction.

Keywords Consciousness · Socially believable behaviour · Conversational agents · Human-machine interaction · Focus tree

M. Gnjatović (✉)
Graduate School of Computer Sciences, John Naisbitt Univeristy, Bulevar Umetnosti 29, 11070 Belgrade, Serbia
e-mail: milangnjatovic@yahoo.com

M. Gnjatović
Faculty of Technical Sciences, University of Novi Sad, Dositeja Obradovića 6, 21000 Novi Sad, Serbia

B. Borovac
Faculty of Technical Sciences, University of Novi Sad,, Trg Dositeja Obradovića 6, 21000 Novi Sad, Serbia
e-mail: borovac@uns.ac.rs

© Springer International Publishing Switzerland 2016
A. Esposito and L.C. Jain (eds.), *Toward Robotic Socially Believable Behaving Systems - Volume II*, Intelligent Systems Reference Library 106, DOI 10.1007/978-3-319-31053-4_4

4.1 Introduction

Considerable research effort has been already devoted to the broad research question of developing cognitive conversational agents [16, 17]. However, at the moment, we as a community are still far away from technical systems that perform socially believable behaviour, and quite rightly. The aim of this paper is to identify some of the causes of the current state in the field, and to propose a way to, at least partially, address them.

Our conceptual approach can be briefly described as follows. We aim beyond cognitive-like technologies—toward conscious-like technologies. More precisely, we propose that socially believable agents should simulate two fundamental features of the high-level phenomenon of human consciousness: the unity of conscious experiences, and the qualitative nature of conscious feelings. Thus, this paper reports on the computational modelling of these features.

The paper is organized as follows. Section 4.2 provides a brief overview of some recent and ongoing research projects related to cognitive technical systems. In Sect. 4.3, we reflect on the often overlooked fact that socially believable behaviour of a technical system is a qualitative requirement, rather than functional. We propose that socially believable agents should, inter alia, simulate fundamental features of human consciousness as a means of fulfilling this requirement. Section 4.4 highlights important pitfalls of the currently dominant approaches to developing cognitive conversational agents. Finally, Sects. 4.5–4.7 introduce and discuss our approach in more details.

4.2 Related Work

It is widely accepted that socially believable systems should be capable of closed loop interaction with humans. This is hardly surprising since interaction has an important role in social relationships. One of the points of departure for research on socially believable technical systems is that the user's reactions to them are evolutionarily conditioned. Two main factors that cause the user to perceive a technical system as intelligent and autonomous are apparent autonomy of robots, and their capacity to engage in a natural language dialogue [46].

Generally speaking, at the level of requirements, the key functionalities of such cognitive systems include perception (e.g., audio and visual recognition, etc.), knowledge representation, interpretation and decision making, learning, planning, natural language processing, affect awareness, etc. Table 4.1 provides an incomplete (but hopefully illustrative) overview of the most prominent functionalities of cognitive technical systems in some recent and ongoing research projects. The list of projects is, of course, incomplete, but the list of cognitive functionalities is illustrative of the state-of-the-art in the field.

Table 4.1 An overview of the most prominent functionalities of cognitive technical systems in some recent and ongoing research projects

Project	Framework*	Perception	Representation	Interpretation, Deciding	Learning	Planning	Language processing	Affect awareness
Assisted Cognition [29]	Uni. Washington	✓	✓	✓	✓	✓		
CALO [33]	DARPA	✓	✓	✓	✓	✓	✓	
RADAR [18]	DARPA	✓	✓	✓	✓	✓	✓	
CoSy [11]	FP6	✓	✓	✓	✓	✓	✓	✓
COGNIRON [12]	FP6	✓	✓	✓	✓		✓	
Companions [53]	FP6	✓	✓	✓			✓	✓
CHIL [49]	FP6	✓	✓	✓			✓	
INHOME [47]	FP6	✓	✓	✓				
SHARE-it [13]	FP6	✓	✓	✓				
humaine-aaac [15]	FP6	✓	✓	✓			✓	✓
SIMILAR [3]	FP6	✓	✓	✓			✓	✓
DIRAC [27]	FP6	✓	✓	✓				
ICEA [55]	FP7	✓	✓	✓	✓	✓		✓
HUMOUR [8]	FP7		✓		✓			
SCANDLE [54]	FP7	✓	✓	✓				
FROG [28]	FP7	✓	✓	✓		✓	✓	✓
Virtual Human [34]	BMBF	✓	✓	✓	✓		✓	✓
SmartKom [48]	BMBF	✓	✓	✓	✓	✓	✓	

(continued)

Table 4.1 (continued)

Project	Framework*	Perception	Representation	Interpretation, Deciding	Learning	Planning	Language processing	Affect awareness
Situated Artificial Communicators [35]	DFG	✓	✓	✓		✓	✓	
Resource-adaptive cognitive processes [42]	DFG	✓	✓	✓	✓		✓	
NIMITEK [51]	SA	✓	✓	✓	✓		✓	✓
A Companion-Technology for Cognitive Technical Systems [50]	DFG	✓	✓	✓	✓	✓	✓	✓
Design of Robots as Assistive Technology for the Treatment of Children with Developmental Disorders [19]	MPNTR	✓	✓	✓	✓		✓	✓

This list of projects is incomplete, but the list of cognitive functionalities is illustrative of the state-of-the-art in the field

*DARPA—Defense Advanced Research Projects Agency; FP6—Sixth Framework Programme; FP7—Seventh Framework Programme; DFG—German Research Foundation; BMBF—German Federal Ministry of Education and Research; SA—Federal State of Saxony-Anhalt; MPNTR—Serbian Ministry of Education, Science and Technological Development

Fig. 4.1 A schematic representation of the closed loop human-machine interaction

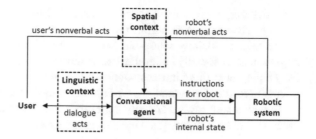

It is evident that at the heart of most of the systems developed within these projects there is a cognitive conversational agent that manages interaction between the user and an external system (e.g., a robot, cf. Fig. 4.1). During the interaction, the user and the system share the linguistic and spatial contexts.[1] The linguistic context includes their dialogue acts, dialogue history, knowledge of the dialogue domain, etc. The spatial context includes positions of the user and the robot, their nonverbal acts, and various information on other spatial entities (animate of inanimate) that are relevant for interaction. Briefly stated, the task of the conversational agent is to interpret the user's dialogue and nonverbal acts in the current interaction context, and to decide whether and how the system as a whole should respond.

However, it should be noted that not all these systems are intended to perform socially believable behaviour. In other words, even a well-designed cognitive architecture of a system does not necessarily imply its social believability. This is discussed in the next section.

4.3 Where to Go from Here?

A great deal of what we refer to as conversational cognitive agents represent software artefacts. In this field, innovative computational conceptualizations and software algorithms constitute the essence of scientific work. Thus, researchers talk predominantly about conceptualizations, models and software algorithms, while software process models are unduly neglected. The primary purpose of software process models is to provide guidance on the order in which a project should carry out its major tasks, and their importance has been widely acknowledged for a relatively long time now [6]. Virtually all software process models recognize the importance of software requirements management. In line with this, some of the major risks to project completion relate to incomplete, inaccurate or vague requirements [31]. Does our field also suffer from such a neglect? Do we omit important development phases or apply them in the wrong order? The answer to these questions is positive.

In a nutshell, when we make decisions related to the design of a system, we focus on the system's functionalities, rather than on the system's features. This is

---

[1] And possibly other contexts (e.g., the user's electroencelographic activity, etc.).

not justifiable, since features cannot be reduced to functionalities. To illustrate this quite clearly: calculators perform so much better than us in calculation, but that does not make them socially believable. Similarly, we should not fail to recognize that the requirement of socially believable behaviour is not functional, but rather qualitative. Developing ever more intricate architectures[2] devoted to support a growing number of cognitive-like functionalities will not necessarily satisfy this requirement.

To take a step toward socially believable conversational agents, we suggest that systems should simulate fundamental features of human consciousness.

4.3.1   The Fundamental Features of Consciousness

The word "consciousness" is used in a variety of ways. Since there is no widely-accepted definition for consciousness, it is important to clarify what we understand under this term. We adopt Searle's view on consciousness [40, 41]. For him, conscious states are higher level features of the brain caused by the lower level neurobiological processes in the brain. But, what is more interesting for the discussion in this paper is his view that consciousness has essential features that make it different from other biological phenomena. These features are: unity and qualitativeness. They are best described in Searle's own words:

- The unity of conscious experiences:

    All conscious experiences at any given point in an agent's life come as part of unified conscious field. If I am sitting at my desk looking out the window, I do not just see the sky above and the brook below shrouded by the trees, and at the same time feel the pressure of my body against the chair, the shirt against my back, and the aftertaste of coffee in my mouth. Rather I experience all of these as part of a single unified conscious field [41, p.41].

- The qualitative nature of conscious feelings:

    Every conscious state has a certain qualitative feel to it […]. The experience of tasting beer is very different from hearing Beethoven's Ninth Symphony, and both of those have a different qualitative character from smelling a rose or seeing a sunset. These examples illustrate the different qualitative features of conscious experiences. One way to put this point is to say that for every conscious experience there is something that it feels like, or something that it is like to have that conscious experience [40, p.39].

Thus, consciousness should not be confused with knowledge, attention or self-consciousness [40, p.8]—and consequently with cognition. So, if we aim at introducing conscious-like features in cognitive technical systems, the beginning of this might be an attempt to provide a computational simulation of these features. This

---

[2]Apart from the fact that the study of architecture to support intelligent behaviour may be one of the most ill-defined enterprises in the field of artificial intelligence [26].

is discussed in the following sections. But, before we elaborate on the underlying computational model in more details, we first discuss two pitfalls that are often encountered in the field.

4.4   Conceptual and Methodological Pitfalls

Two main pitfalls of the currently dominant approaches to developing cognitive conversational agents may be summarized as follows:

- dogmatic application of purely statistical approaches (cf. also [19, 21]),
- misinterpretation of the notion of dialogue structure (cf. also [24]).

We discuss them in the rest of this section.

4.4.1   The Methodological Pitfall of Purely Statistical Approaches

The debate on the mechanisms underlying language acquisition has settled into two main directions of research. The first, represented by researchers as Chomsky, suggests that the ability to use language is innate, i.e., a part of our biological endowment [9]. The second, represented by researchers as Saffran and Tomasello, suggests learning-oriented theories, e.g., that learners, including infants, use statistical properties of linguistic input [37] and the inherited cognitive skills of intention-reading and pattern-finding to acquire language [45]. This intellectual divide is reflected in the field of human-machine dialogue. The first research direction relates to studying and modeling the mechanisms underlying the human language processing system (e.g., attention, memory, perception, etc.) in order to computationally simulate their features. The second, prevalent research direction is primarily intended to simulate language behavior without a deep understanding of the underlying phenomena. The principal approach is to use data derived from language corpora, and apply automated analysis methods to empirically derive rules and structures to manage machine dialogues [52]. These statistical methods analyze only external language behavior (i.e., corpora), and do not account for innate features of the human language processing system.

Statistical approaches have proved to be successful when applied in certain aspects of machine learning, such as automated speech recognition or machine translation. Following this methodological trend, significant research efforts are currently devoted to apply statistical approaches to address the more general research question of managing human-machine dialogues that require some sort of contextual analysis and interpretation. However, the last two decades have shown that they are insufficient to address this research question. The reason lies in the data-driven methodology with no explanation power. Corpus design criteria usually correspond with intuitive, and hence unreliable, judgments of the researcher [43, 44, p.91]. It cannot be expected that language corpora, no matter how carefully produced, will contain surface manifestations of all relevant dialogue phenomena. Therefore, constructing computational models based solely on a statistical analysis of language corpora cannot adequately address the research question of managing human-machine dialogues. To address this research question, we need to involve linguistic structures that underlay the human language processing system and cannot be extracted from language corpora. The point of departure for our approach is that insights in neuroimaging studies may bring us a step closer to identifying and understanding these structures and related cognitive processes.

4.4.2   The Conceptual Pitfall of Misinterpreting the Dialogue Structure

The research question of modelling human-machine dialogue structure is still not satisfactorily answered. One of the most obvious principle in conversation is that each dialogue act creates a possibilities of appropriate response dialogue acts [39, p.8]. This principle of linguistic exchange is observable in Schegloff's [38] "adjacency pairs" and Roulet's dialogue "moves". An adjacency pair represents a simple linguistic exchange containing two sequent and pragmatically related dialogue acts, such as question/answer, greeting/greeting, offer/acceptance, etc. In contrast to the sequential structure of adjacency pairs, Roulet's suggests that linguistic exchange has a hierarchical structure. He observes relations in conversation between so-called moves each of which may contain several dialogue acts. For Roulet [36, p.94], the intentionality shared between interlocutors is a constant object of negotiation. This activity of negotiation determines the structure of linguistic exchange—any linguistic exchange consists of at least three phases: proposition, reaction and evaluation. In addition, linguistic exchange may be developed recursively. To illustrate this, an interlocutor may provide a reaction only if the proposition is clear and complete. Otherwise, he may open a secondary linguistic exchange in order to get the additional information. Upon successful closure of the secondary linguistic exchange, both participants can get back to the primary linguistic exchange. Thus, the concept of recursivity in the development of linguistic exchange implies a hierarchical structure of conversation (cf. Fig. 4.2). Similarly, Grosz and Sidner [25] propose a hierarchical structure of intention in conversation.

The understanding that conversation has a hierarchical structure has influenced, directly or indirectly, many computational models of human-machine dialogue. However, it appears that this linguistic insight was misinterpreted in an unintended way (cf. also [25, pp.180–2]). To illustrate this, we consider two spoken dialogue systems: the SUNDIAL system designed to handle travel conversations [4], and the speech translation system Verbmobil [1]. The dialogue structure underlying the implementation of the SUNDIAL system is defined as a hierarchy of four levels: the transaction

Fig. 4.2 Hierarchical
structure of the linguistic
exchange (adjusted from [36,
p.97])

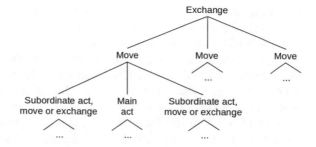

level, the exchange level, the intervention level (i.e., conversation units that carry illo-
cutionary functions of initiative, reaction, or evaluation), and the dialogue act level.
Quite similarly, in the Verbmobil system, the dialogue structure is divided in the
following four levels: the dialogue level, the phase level (distinguishing the phases
of greeting, negotiation, closing), the turn level (the main turn classes for negotiation
dialogues are: initiative, response, confirmation, etc.), and the dialogue act level.

They claim to model the dialogue structure for a specific domain of interaction,
but in fact, they model the task structure. In both cases, the task structure is preset,
and a task-specific role is assigned to each dialogue constituent (e.g., initiative,
response, confirmation, etc.). At run time, these dialogue managers take an ensuing
user's dialogue act and try to assign one of predefined roles to it—i.e., to map user's
dialogue acts onto a set of predefined dialogue constituents.

What is wrong with this solution? First, the task structure is not the same as
the dialogue structure. The task structure is a special case of the intentional structure, which is again just one of the components of the dialogue structure (cf. [25]).
Thus, the reduction of the dialogue structure to the task structure is not justifiable.
In practice, the determination of an ensuing dialogue act in terms of how well it
matches a predefined task-specific role implies a high level of task-orientation of
the observed systems. It is interesting to note that both these approaches consider
telephone conversations. Telephone conversations are specific because they inherently have well defined objectives [39], and thus they are easier to model in terms
of task-specific constraints. However, this certainly does not hold for more general
cases of conversation.

Second, the task structures in task-oriented dialogues are not necessarily prebuilt
[25, p.182]. It has been widely acknowledged that intentionality is not given at the
beginning of a conversation, but evolves as the conversation proceeds [25, 36, 39].
Being a special case of the intentional structure, the same holds for the task structure.

4.4.3 Addressing the Pitfalls

To address the aforementioned pitfalls, in previous work we have introduced the
focus tree model [23, 24]. It is a representational and cognitively-inspired model

of attentional information in machine dialogue that addresses the research problems of robust interpretation of different syntactic forms of spontaneously uttered users' commands with no explicit syntactic expectations, and flexible designing of adaptive dialogue strategies. The level of detail of the model is sufficient for a computational implementation, while the level of abstraction is sufficient to enable its generalization over different interaction domains. Thus, the model was successfully applied in several prototypical dialogue systems with diverse domains of interaction. For a more detailed insight into the focus tree model, the reader may consult [19–24]. In the following sections, we show how this model can be applied to simulate certain aspects of the fundamental features of consciousness.

4.5 Unified Interpretation

The focus tree is a hierarchical structure that represents the activated part of the long-term memory. Each node of the focus tree represents a semantic entity within the observed interaction domain. The focus of attention summarizes information from previous interaction (i.e., a history of the interaction), and at any moment it is placed on exactly one node. When the focus of attention is placed on node n, it signals that semantic entities represented by n and its ancestors in the focus tree are currently retrieved in working memory. In other words, the top-down path starting at the root node and ending at n represents the currently activated mental representation [23, 24].

In general, each top-down path starting at the root node and ending at an inner of a terminal node represents a mental representation. The relationship between nodes may be shortly summarized as follows: if q is a descendant of p, it means that the mental representation ending at q encapsulates and extends the mental representation ending at p. To illustrate this, let us consider a very simple interaction domain containing two entities: a robot and a glass. The robot can use its arm to point to the glass, or to move it in one of two directions (leftward and rightward). In addition, the robot's head has an eye system, so it can be instructed to close or open its eyes, to blink, or to direct its gaze in one of two directions (leftward and rightward). A focus tree that represents this interaction domain is depicted in Fig. 4.3.

The processing aspect of this model addresses the research question of mapping propositional content of the user's dialogue act onto the focus tree. One of the most obvious features of the processing algorithm is that it does not explicitly include syntactic expectations. Although this decision is in line with the primary intention to enable conversational agents to interpret arbitrary syntactic forms of spontaneously uttered users' dialogue acts, it may appear too restrictive. However, it is not necessarily a disadvantage of the model.

It should be noted that the model is in line with the Chomsky's concept of the kernel of language [10, p.45,61]. For him, the kernel of language consists of a finite number of simple, declarative, active sentences whose terminal strings are derived by a simple system of phrase structure. In addition, every sentence that does not

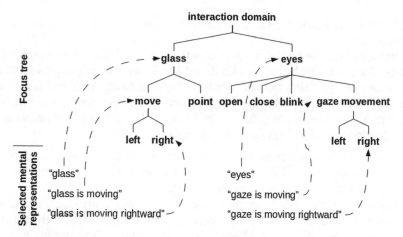

Fig. 4.3 A focus tree representing the observed interaction domain

belong to the kernel can be derived from the strings underlying kernel sentences by a sequence of one or more transformations. Following this insight, the focus tree can be interpreted as comprising the kernel sentences within the observed dialogue domain. Each top-down path starting at the root node represents a kernel sentence, and every dialogue act within the observed dialogue domain can be mapped to one or more kernel sentences.

So, except for the canonical syntactic structures, we give up on syntax. We intentionally conceptualize the user's dialogue acts as sets of keywords and nonrecursive phrases that carry propositional content, with no further syntactic specification. Thus, the interpretation of a dialogue act is based on the mapping of a set of propositional *symbols* (i.e., keywords and phrases referring to semantic entities within the dialogue domain) onto the focus tree. It is particularly important to note that these symbols are not tied to a particular perceptual system (e.g., linguistic), but that they are *amodal*. This allows us to apply the focus tree model to interpret stimuli that are not only linguistic, but also visual, haptic, etc.—without the need to change the processing algorithm.

Consequently, this trade-off between syntax and amodal representation allows for *unified interpretation* of inputs of various communication modalities. To illustrate this, let us consider the following three interaction acts of the user:

- a single verbal command: "please move the glass leftward",
- a sequence of two verbal commands, "please take the glass" and "move it leftward",
- a composition of verbal command "move it leftward" and nonverbal action of pointing to the glass.

To perceive the first two interaction acts, a system needs a speech recognition module. For the third interaction act, it needs a visual recognition module as well. However, apart from using different sets of the perception modules, the perception result for all these interaction acts is the same—a set of amodal propositional symbols that relate to the following semantic entities: *the glass, the action of moving, the leftward direction*.

The task of the interaction management module is to interpret the perceived set of propositional symbols, i.e., to map it onto the focus tree. It is important to note that for the interpretation of these symbols it is completely irrelevant from which perception module a particular symbol is coming. Before we discuss the processing algorithm in more details, we should clarify the conceptualization of propositional symbols. These symbols may be classified in two groups[3]:

- Stimuli—symbols that relate to semantic entities within the interaction domain that should be retrieved in working memory to interpret a given interaction act. All symbols from the above examples are stimuli.
- Inhibitors—symbols that relate to semantic entities within the interaction domain that must not be retrieved in working memory while interpreting a given interaction act. For example, dialogue act "do not move" contains an inhibitor implying that the action of moving should not be included in the resulting mental representation.

This classification of propositional symbols is inspired by two related concepts that have been widely acknowledged in cognitive psychology—the concepts of mnemonic selection and cognitive inhibition. The idea underlying the concept of mnemonic selection is that representations held in (working) memory can be prioritized according to their momentary relevance to ongoing cognitive operations [5, pp.172–3]. The idea underlying the concept of cognitive inhibition is that mental representations that are not relevant for ongoing cognitive operations can be inhibited (i.e., stopped or overridden) [32, pp.4–5]. With respect to the interpretation of interaction acts, these concepts reflect the fact that not all semantic entities within the interaction domain are of equal importance all the time. A fundamental aspect of the functionality of a conversational agent refers to the ability to dynamically select relevant semantic entities and suppress irrelevant semantic entities. In our approach, stimuli and inhibitors provide information to distinguish between relevant and irrelevant semantic entities in a given moment of interaction.

In previous work, we have considered the interpretation of verbal dialogue acts in algorithmic terms [24], and in terms of the basic cognitive functions that provide means for the flexibility of working memory, i.e., mnemonic selection, updating the focus of attention, and updating the contents of working memory [23]. Here, we consider the essential part of the algorithm and generalize it for multimodal user's acts.

For the purpose of easier representation in this and the next section, we use the following abbreviations:

- S—a set of stimuli perceived from the current user's interaction act,
- I—a set of inhibitors perceived from the current user's interaction act,
- n—a node of the focus tree,

---

[3]In our previous work, we refer to these two groups as to *focus stimuli* and *negative reinforcement stimuli*. However, the terms *stimuli* and *inhibitors* reflect more appropriately the dichotomy between these groups.

- K—a set of all top-down paths starting at the root node and ending at a terminal node,
- $A(n)$—a set that contains node n and all of its ancestors in the focus tree,
- $T(n)$—a set that contains all terminal nodes that are descendants of node n (if n is a terminal node, then $T(n) = \{n\}$),
- $\hat{D}$—the number of all nodes in the focus tree,
- $\hat{T}$—the number of all terminal nodes in the focus tree.

Isolated from the interaction context, every top-down path whose nodes cover all stimuli from S, and do not cover inhibitors from I represents a possible interpretation of the current user's interaction act. Let P_{INT} be a set containing all these paths:

$$P_{INT} = \{P \mid (P \in K) \wedge (S \subset P) \wedge (I \cap P) = \emptyset)\} . \qquad (4.1)$$

If this set is not empty, it means that the user's interaction act can be interpreted within a given interaction domain. However, these interpretations are not equally appropriate. To include the context in the interpretation, we take the current focus of attention n_{FA} into account. We recall that the set of nodes on the top-down path starting at the root node and ending at n_{FA} represents the currently activated mental representation. Thus, the most appropriate interpretation can be described as the one that contains the largest number of the nodes from the currently activated mental representation. Still, in P_{INT} there can be more interpretations containing the same maximum number of nodes from the currently activated mental representation. Let P_{CAND} be a set containing these interpretations:

$$P_{CAND} = \{P \mid (P \in P_{INT}) \wedge (|P \cap A(n_{FA})| = \max_{P_i \in P_{INT}} |P_i \cap A(n_{FA})|)\} . \qquad (4.2)$$

P_{CAND} cannot be empty, since all interpretations contain the root node. If P_{CAND} contains only one interpretation, it is chosen to be the new activated mental representation. In this case, we say that the user's interaction act is *complete*. Otherwise, the new activated metal representation is defined as the intersection of all interpretations in P_{CAND}:

$$M_R = \begin{cases} P \mid P \in P_{CAND} & \text{if } |P_{CAND}| = 1, \\ \bigcap_{P_i \in P_{CAND}} P_i & \text{if } |P_{CAND}| > 1. \end{cases} \qquad (4.3)$$

In both cases, the focus of attention is updated accordingly. The new focus of attention, $n_{FA}^\star$, is the lowest node from M_R. If the new focus of attention is the same as the previous focus of attention, i.e., $n_{FA}^\star, = n_{FA}$, we say that the user's act is *ambivalent*, otherwise it is *incomplete*.

The case that remains to be considered is when P_{INT} is empty. That implies that the user's interaction act cannot be interpreted in the observed interaction domain. We refer to such an act as to *semantically incorrect*. We recall that the user's interaction act is represented by a set of propositional symbols (i.e., stimuli and inhibitors) which is a union of symbols perceived by different perception modules (e.g., for speech and visual recognition). The semantic incorrectness of an act may be caused by the fact that the recognition results of different perceptions modules are conflicting. To resolve this situation, the system tries to separately interpret propositional symbols perceived by each perception module. The interpretation steps are the same as described above. If only one of this separate inputs can be interpret, the system takes this interpretation to be the new activated mental representation. If there is more than one input that can be interpret, the system chooses the interpretation that contains the largest number of the nodes from the currently activated mental representation.

To illustrate the introduced algorithm, we consider four examples of interpretation of the user's interaction acts, summarized in Table 4.2 and depicted in Fig. 4.4.

In the first example, the user's interaction act contains two stimuli—glass, and the action of pointing. There is only one top-down path in the focus tree that contains

Table 4.2 Four examples of interpretation of the user's interaction acts

Node carrying the current focus of attention (n_{FA})	Perceived stimuli	Node carrying the next focus of attention (n_{FA}^*)	Classification
(Any node)	Glass, the action of pointing	Point	Complete
Interaction domain	Glass, the action of moving	Move	Incomplete
Interaction domain	The action of moving	Interaction domain	Ambiguous
Eyes	The action of moving	Gaze movement	Incomplete

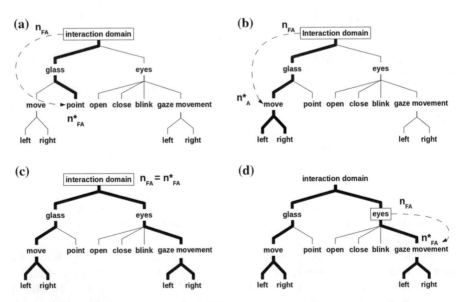

Fig. 4.4 Illustrations of the context-dependent interpretation of the user's interaction acts. The *arrows* denote the transition of the focus of attention

nodes related to these stimuli (cf. Fig. 4.4a). Thus, this interpretation becomes the activated mental representation, the focus of interaction is placed on node *point*, and the act is classified as complete. It should be noted that for the interpretation of a complete command it is not important on which node the focus of attention was previously placed.

In the second example, the user's interaction act contains two stimuli—glass, and the action of moving. There are two top-down paths that contain nodes related to these stimuli (cf. Fig. 4.4b). To evaluate the appropriateness of these interpretations, the current focus of attention must be taken into account. Let us assume that the focus of attention is placed on the root node. It means that the currently activated mental representation comprises only this node. According to the algorithm, the interpretations are equally appropriate, since they both encapsulate the currently active mental representation. In addition, the newly activated mental representation is defined as the intersection of these interpretations, which means that the focus of attention is transited to node *move*, and the act is classifies as incomplete.

Let us now consider the user's interaction act that can be reduced to only one stimulus—the action of moving. In the given focus tree, there are four top-down paths that contain a node representing the action of moving (as indicated in Fig. 4.4c, d). both the glass and the eyes of the robot can be moved leftward or rightward. However, the interpretation of this act depends on the current focus of attention. If the current focus of attention is placed on the root node (Fig. 4.4c), all four possible interpretations of the user's interaction act are equally appropriate (similarly as in the previous case), and the newly activated mental representation is defined as the intersection of these paths, which in this case means that the focus of attention remains on the same node. The user's interaction act is classifies as ambiguous.

Otherwise, if the current focus of attention is placed on node *eyes* (Fig. 4.4d), this means that the currently activated mental representation contains nodes *interaction domain* and *eyes*. Now, according to the introduced algorithm, the interpretations containing these nodes are preferred. The new activated mental representation is defined as the intersection of these interpretations, and the new focus of attention is placed on node *gaze movement*. Consequently, this interaction act is classified as incomplete.

The classification of the user's interaction acts (i.e., complete, incomplete, ambiguous, and semantically incorrect) brings us closer to a conceptualization of the interaction act quality. In this particular view, the quality of an interaction act is related to its interpretation complexity. The next section discusses this conceptualization in more details.

4.6   Qualitativeness

A logical way to conceptualize the notion of qualitativeness in the interaction context is to relate it to interaction acts and states. In cognitive terms, we relate the "quality" of an interaction act to the estimated cognitive load required to interpret it. Since an interaction act is always processed in a contextual framework, the cognitive load is a context-dependent measure—the interpretation of an interaction act depends not only on the act itself, but also on the content of working memory immediately before the act has been interpreted (i.e., the activated mental representation).

To illustrate this, we refer to the notion of event related potential (ERP), i.e., the voltage fluctuations across neural membranes in the brain that are time-locked to a sensory, motor or cognitive stimulus [30, p. 464]. We consider two particular ERPs that reflect real-time language comprehension:

- The N400 is a negative deflection of the ERP signal peaking at about 400 ms after stimulus-onset. Its amplitude can be interpreted in terms of memory retrieval, i.e., the N400 amplitude reflects the mental processes that accompany the retrieval of lexical information from long-term memory [7, pp.128,134,139].
- The P600 is a positive deflection of the ERP signal that reaches maximum around 600 ms after stimulus-onset. Its amplitude can be interpreted in terms of semantic integration, i.e., the P600 amplitude reflects the integration of the activated lexical information into the existing current mental representation of an unfolding sentence. Integration difficulty is determined by how much the current mental representation needs to be adapted to incorporate the current input [7, pp.128,133,138].

In technical terms, the "quality" of an interaction act is a measure of the complexity of its context-dependent interpretation. Following the functional interpretations of the N400 and P600, we introduce two parameters of the dialogue act complexity:

- the retrieval cost—simulating the value of the N400 amplitude,
- the integration cost—simulating the value of the P600 amplitude.

Before we formally define these parameters, we recall the interpretation of the focus of attention in our approach. If the focus of attention is placed on node n_{FA}, it means that all semantic entities represented by n_{FA} and its ancestors in the focus tree are currently retrieved in working memory. This set of node is denoted by $A(n_{FA})$. Isolated from the interaction context, the retrieval cost of this particular interaction state is proportional to the number of elements in $A(n_{FA})$ (Fig. 4.5a). However, we should take the interaction context into account. If the focus of attention was placed on node n_{FA}, and after the interpretation of an interaction act transited to node $n_{FA}^{\star}$, then the set difference between $A(n_{FA}^{\star})$ and $A(n_{FA})$ (i.e., set $A(n_{FA}^{\star}) \setminus A(n_{FA})$) represents semantic entities that were retrieved in the working memory during the interpretation (Fig. 4.5b, c and d). To describe the general case, let us assume the focus of attention was placed on node n_{FA}, and that the interpretation of an interaction act results in a set of focus candidate nodes $N = \{n_1, n_2, \ldots, n_k\}$, where $k \geq 0$. Semantic entities that have been retrieved during the interpretation are represented by the set difference between (i) the union of all $A(n)$, where $n \in N$, and (ii) $A(n_{FA})$. The retrieval cost of the observed interaction act is proportional to the number of elements in the resulting set. This number can be further normalized by dividing it by the number of all nodes in the focus tree (which we denote by $\hat{D}$). Finally, the retrieval cost of the interpretation of an interaction act is defined as follows:

$$\rho = \begin{cases} \dfrac{1}{\hat{D}} & , |N| = 0 \\ \dfrac{|(\bigcup\limits_{n \in N} A(n)) \setminus A(n_{FA})|}{\hat{D}} & , \text{otherwise.} \end{cases} \tag{4.4}$$

The integration cost is related to the number of terminal nodes. The number of terminal nodes that are descendants of n_{FA} (this set is denoted by $T(n_{FA})$) is the same as the number of top-down paths (i.e., complete interpretations) that contain n_{FA}. Isolated from the interaction context, the integration cost of this particular interaction state is proportional to the number of elements in $T(n_{FA})$ (Fig. 4.5). This number can be further normalized by dividing it by the number of all terminal nodes in the focus tree (which we denote by $\hat{T}$). In general, the integration cost of the interpretation of an interaction act is defined as follows:

$$\iota = \begin{cases} 1 & , |N| = 0 \\ 0 & , (|N| = 1) \wedge (\text{n is terminal node} \mid n \in N) \\ \dfrac{|\bigcup\limits_{n \in N} T(n)|}{\hat{T}} & , \text{otherwise.} \end{cases} \tag{4.5}$$

The processing costs of an interaction act always lie within the range [0, 1]. Thus, for a semantically irregular interaction act, both the retrieval and integration costs have the maximum value of one. For an complete interaction act, the integration cost has the minimum value of zero. The processing costs are elaborated in more details in [22]. The next section discusses how they can facilitate context-dependent reasoning.

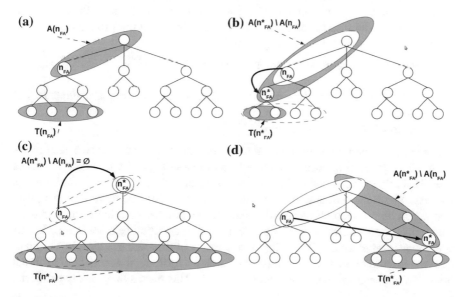

Fig. 4.5 Illustrations of the retrieval and integration costs. The *arrows* denote the transition of the focus of attention

4.7   Discussion: Ongoing and Future Work

We use the introduced processing costs to address two important research questions in the context of human-interaction: the system's adaptation in long-term interaction,[4] and context-dependent perception. This section reports on our ongoing and future work on these questions.

4.7.1   Learning Through Interaction

The research problem of automatic construction of a focus tree representing a given interaction domain can be divided in two tasks:

- identification of the set of semantic entities within the interaction domain (i.e., nodes).
- identification of the parent-child relationships between nodes (i.e., edges).

The first task can be reduced to information retrieval and accomplished by applying the well-elaborated techniques, which we do not consider in this paper. Instead, we focus on the second task.

The evaluation of the processing costs of an interaction act can be generalized to interaction fragments, i.e., a sequence of chronologically ordered interaction acts. Thus, the retrieval cost of an interaction fragment can be defined as a sum of the retrieval costs of each interaction act comprised in the given fragment. Similarly, the integration cost of an interaction fragment is defined as a sum of the integration costs of comprised interaction acts. It should be noted that the processing costs of an interaction fragment depend on several factors:

- the propositional content of comprised interaction acts (i.e., perceived propositional symbols),
- the chronological order of interaction acts (i.e., a sequence of intermediate foci of attention),
- and the topology of the focus tree.

However, for a particular observed interaction fragment, the first two factors are fixed, so its processing costs depend only on the topology of the underlying focus tree. This allows for a corpus-based evaluation of a focus tree. And furthermore, the research problem of automatic construction of a focus tree representing a given interaction domain may be reduced to the problem of constructing a focus tree for which the processing costs of the given corpus of interaction acts are minimal. This is elaborated in more details in [20]. Two aspects of this research problem are particularly important. First, according to the Cayley's formula, the number of different trees that can be constructed on n nodes is $n^{(n-2)}$. This means that for any nontrivial

---

[4]The research question of short-term adaptation of the system's dialogue strategy is addressed in [23, 24].

interaction domain, it is not feasible to construct and evaluate all possible trees in a reasonable time. Therefore, different optimization techniques will be developed and investigated. Second, the processing cost contains two parameters (the retrieval and integration costs) which are not equally descriptive in different interaction contexts. Thus, a number of different combinations of these parameters will be explored.

One of the requirements for long-term human-machine interaction is that the machine can adapt to different human interlocutors, or to changes of the interaction style of an interlocutor. In our approach, the system keeps track on the interaction history,[5] and collects a corpus of the user's interaction acts. Thus, the underlying focus tree may be constantly evaluated and, when appropriate, modified with respect to the corpus. This representational corpus-based approach underpins our ongoing work on machine learning through interaction.

4.7.2   Context-Dependent Perception

The importance of contextual information in speech and visual recognition has been widely acknowledged. At the methodological level, the state-of-the-art approaches to the research question of context-dependent recognition usually rely on statistical post-processing of recognition hypotheses [2, 14]. In contrast to these approaches, we apply a representational approach based on the processing costs of interaction acts. Here we very briefly illustrate the main idea underlying this approach.

Keeping in mind that the retrieval and integration costs of an interaction act depends, inter alia, on the current focus of attention, one of the criteria to compare recognition hypotheses is with respect to their context-dependent processing complexity. Thus, interaction acts of lower complexity could be considered to be more likely than interaction acts of higher complexity. Another criterion to compare recognition hypotheses is with respect to the interlocutor's interaction style, reflected through the processing costs of his interaction acts. For example, if the retrieval costs of several subsequent interaction acts are relatively small, and the integration costs are decreasing, it may signal that the interlocutor formulates propositional content in a gradual and highly contextualized manner (i.e., he produces a sequence of incomplete interaction acts which, when observed together, have a complete propositional meaning). Or, if the retrieval costs of several subsequent interaction acts are significant,[6] and the integration costs are small (or equal to zero), it may signal that the interlocutor shifts between different interaction topics, but he tends to produce complete interaction acts. As part of future work, different conceptualizations of interaction style will be introduced and investigated. In general, a conversational

---

[5]The practice of long-term collection of the user's data opens many important ethical questions that are out of the scope of this paper. However, these questions still remain to be properly addressed.

[6]Only for the purpose of illustration, the lower end of the range of significant retrieval costs for a given balanced focus tree can be estimated as the ratio of the height of the focus tree to the number of all nodes in the focus tree.

agent will be able to estimate the interlocutor's interaction style. Consequently, a recognition hypothesis that is more suited to the interlocutor's style could be preferred.

4.8 Conclusion

This paper summarized our previous work and highlighted some of our ongoing work on socially believable conversational agents. We considered the capability of a technical system to simulate fundamental features of the high-level phenomenon of human consciousness as one of the key desiderata to perform socially believable behaviour. It is important to emphasize that this capability goes beyond other cognitive-like technical functionalities (e.g., perception, interpretation, deciding, etc.) often mentioned in the literature. It rather unifies and builds upon them. This is in line with our understanding that socially believable behaviour of a technical system is a qualitative, not functional, requirement.

Conscious experiences are always subjective, since they can exist only as experienced by some subject [40]. In our approach, the main idea underlying the conceptualization of subjectivity of conscious experiences in the interaction context may be briefly described in the following points: (i) A focus tree represents a subjective knowledge about the interaction domain. It is important to note that different conversational agents may use different focus trees to represent the same interaction domain shared between them. For example, these focus trees may include the same or similar sets of nodes arranged in significantly different topologies. In other words, these trees encapsulate the same or similar sets of semantic entities from the interaction domain arranged in different meaning representations (conceptualized as top-down paths). (ii) The interpretation of users' interaction acts is subjective. The perceived stimuli and inhibitors are interpreted with respect to the given focus tree and the current focus of attention. It implies not only that the same set of perceived stimuli may be interpreted differently by different conversational agents, but also that even the same conversational agent may interpret differently the same set of stimuli at different times (depending on the current focus of attention).

Finally, it is fair to say that we are still a long way from developing socially believable conversational agents. But one thing is for sure—if we aim at a substantial progress, we need to reconsider the currently prevalent statistical approaches, and investigate alternative ways to conceptualize and model human-machine interaction. This paper is an attempt to take a step in this direction.

Acknowledgments The presented study was sponsored by the Ministry of Education, Science and Technological Development of the Republic of Serbia under the Research grants III44008 and TR32035. The responsibility for the content of this paper lies with the authors.

References

1. Alexandersson J, Reithinger N (1997) Learning dialogue structures from a corpus. In: Proceedings of EuroSpeech-97, Rhodes, pp 2231–2235
2. Bacchiani M, Rybach D (2014) Context dependent state tying for speech recognition using deep neural network acoustic models. In: IEEE international conference on acoustics, speech and signal processing (ICASSP 2014), Florence, pp 230–234
3. Bernsen NO, Dybkjær L (2010) Multimodal usability. Springer, London
4. Bilange E (2000) In: Néel F, Bouwhuis DG, Taylor MM (eds) The structure of multimodal dialoque II. An approach to oral dialogue modelling. John Benjamins Publishing Company, Amsterdam
5. Bledowski C, Kaiser J, Rahm B (2010) Basic operations in working memory: contributions from functional imaging studies. Behav Brain Res 214(2):172–179
6. Boehm BW (1988) A spiral model of software development and enhancement. Computer 21(5):61–72
7. Brouwer H, Fitz H, Hoeks J (2012) Getting real about semantic illusions: rethinking the functional role of the P600 in language comprehension. Brain Res 1446:127–143
8. Casadio M, Giannoni P, Morasso PG, Sanguineti V (2009) A proof of concept study for the integration of robot therapy with physiotherapy in the treatment of stroke patients. Clin Rehabil 23(3):217–228
9. Chomsky N (2000) New Horizons in the study of language and mind. Cambridge University Press, Cambridge
10. Chomsky N (2002) Syntactic Structures. Walter De Gruyter
11. Christensen HI, Sloman A, Kruijff G-J, Wyatt JL (2010) Cognitive systems introduction. In: Christensen HI, Kruijff G-J, Wyatt JL (eds) Cognitive systems. Springer, Berlin, pp 3–50
12. Clodic A, Alami R, Montreuil V, Li S, Wrede B, Swadzba A (2007) A study of interaction between dialog and decision for human-robot collaborative task achievement. In: Proceedings of the 16th IEEE international symposium on robot and human interactive communication, RO-MAN 2007, Jeju Island, Korea, pp 913–918
13. Cortés U, Annicchiarico R, Urdiales C, Barrué C, Martínez A, Villar A, Caltagirone C (2008) Supported Human Autonomy for Recovery and Enhancement of Cognitive and Motor Abilities Using Agent Technologies. Agent Technology and e-Health. Whitestein Series in Software Agent Technologies and Autonomic Computing, Birkhäuser Basel, pp 117–140
14. Dahl G, Yu D, Deng L, Acero A (2012) Context-dependent pre-trained deep neural networks for large vocabulary speech recognition. IEEE Trans Audio Speech Lang Process 20(1):30–42
15. Douglas-Cowie E, Cowie R, Sneddon I, Cox C, Lowry O, McRorie M, Martin JC, Devillers L, Batliner A (2007) The humaine database: addressing the needs of the affective. In: Paiva A, Prada R, Picard R (eds) 2nd international conference on affective computing and intelligent interaction (ACII', 2007), LNCS, 4738. Lisbon, Portugal, pp 488–500
16. Esposito A, Vinciarelli A, Haykin S, Hussain A, Faundez-Zanuy M (2011) Cognitive behavioural systems. Cognit Comput 3(3):417–418
17. Esposito A, Fortunati L, Lugano G (2014) Modeling emotion, behavior and context in socially believable robots and ICT interfaces. Cognit Comput 6(4):623–627
18. Garlan D, Schmerl B (2007) The RADAR architecture for personal cognitive assistance. Int J Softw Eng Knowl Eng 17(2):171–190
19. Gnjatović M (2014) Therapist-centered design of a robot's dialogue behavior. Cognit Comput 6(4):775–788
20. Gnjatović M (2014) Changing concepts of machine dialogue management. In: Proceedings of the 5th IEEE international conference on cognitive infocommunications, cogInfoCom 2014, Vietri sul Mare, Italy, pp 367–372
21. Gnjatović M, Delić V (2012) A cognitively-inspired method for meaning representation in dialogue systems. In: Proceedings of the 3th IEEE international conference on cognitive infocommunications, cogInfoCom 2012, Košice, Slovakia, pp 383–388

22. Gnjatović M, Delić V (2013) Electrophysiologically-inspired evaluation of dialogue act complexity. In: Proceedings of the 4th IEEE international conference on cognitive infocommunications, cogInfoCom 2013, Budapest, Hungary, pp 167–172
23. Gnjatović M, Delić V (2014) Cognitively-inspired representational approach to meaning in machine dialogue. Knowl-Based Syst 71:25–33
24. Gnjatović M, Janev M, Delić V (2012) Focus tree: modeling attentional information in task-oriented human-machine interaction. Appl Intell 37(3):305–320
25. Grosz B, Sidner C (1986) Attention, intentions, and the structure of discourse. Comput Linguist 12(3):175–204
26. Hawes N, Wyatt JL, Sridharan M, Jacobsson H, Dearden R, Sloman A, Kruijff G-J (2010) Architecture and representations. In: Christensen HI, Kruijff G-J, Wyatt JL (eds) Cognit Syst. Springer, Berlin, pp 51–93
27. Hermansky H (2006) A Brief description of DIRAC project. In: Proceedings of the euCognition inaugural meeting: the European network for the advancement of artificial cognitive systems, nice, France
28. Karreman DE, van Dijk EMAG, Evers V (2012) Using the visitor experiences for mapping the possibilities of implementing a robotic guide in outdoor sites. In: Proceedings of the 21st IEEE international symposium on robot and human interactive communication, RO-MAN 2012. France, Paris, pp 1059–1065
29. Kautz H, Fox D, Etzioni O, Borriello G, Arnstein L (2002) An Overview of the Assisted Cognition Project. In: Proc. of the AAAI (2002) workshop on Automation as Caregiver: The Role of Intelligent Technology in Elder Care. Edmonton, Alberta, Canada, pp 60–65
30. Kutas M, Federmeier KD (2000) Electrophysiology reveals semantic memory use in language comprehension. Trends Cognit Sci 4(12):463–470
31. Leishman TR, Cook DA (2002) Requirements Risks Can Drown Software Projects. CrossTalk: The Journal of Defense Software Engineering (April 2002), pp 4–8
32. MacLeod CM (2007) The concept of inhibition in cognition. In: Gorfein DS, MacLeod CM (eds) Inhibition in cognition. American Psychological Association, Washington, pp 3–23
33. Maheswaran R, Tambe M, Varakantham P, Myers K (2004) Adjustable autonomy challenges in personal assistant agents: a position paper. In: Nickles M, Rovatsos M, Weiss G (eds) Agents and computational autonomy. Lecture notes in computer science, vol 2969. Springer, Berlin, pp 187–194
34. Reithinger N, Gebhard P, Löckelt M, Ndiaye A, Pfleger N, Klesen M (2006) VirtualHuman—dialogic and emotional interaction with virtual characters. In: Proceedings of the eighth international conference on multimodal interfaces, (ICMI, 2006), Banff, Canada
35. Rickheit G, Wachsmuth I (1996) Collaborative research centre situated artificial communicators at the University of Bielefeld, Germany. In: Mc Kevitt P (ed) Integration of natural language and vision processing, Springer, Netherlands, pp 165–170
36. Roulet E (1992) On the structure of conversation as negotiation. In: Parret H, Verschueren J (eds) (On) Searle on conversation. John Benjamins Publishing Company, Amsterdam, pp 91–99
37. Saffran JR (2003) Statistical language learning: mechanisms and constraints. Curr Direct Psychol Sci 12(4):110–114
38. Schegloff EA (1968) Sequencing in conversational openings. Am Anthropol 70:1075–1095
39. Searle J (1992) Conversation. In: Parret H, Verschueren J (eds) (On) Searle on conversation. John Benjamins Publishing Company, Amsterdam, pp 7–29
40. Searle J (2002) Consciousness. Consciousness and language. Cambridge University Press, Cambridge
41. Searle J (2002) The problem of consciousness. Consciousness and language. Cambridge University Press, Cambridge
42. Siekmann J, Crocker MW (2010) Resource-Adaptive Cognitive Processes. In: Crocker MW, Siekmann J (eds) Resource-adaptive cognitive processes. Springer-Verlag, Berlin Heidelberg, pp 1–10

43. Sinclair J (2005) Corpus and text—basic principles. In: Wynne M (ed) Developing linguistic corpora: a guide to good practice. Oxbow Books, Oxford, pp 1–16
44. Tognini-Bonelli E (2001) Corpus linguistics at work. John Benjamins, Amsterdam
45. Tomasello M (2003) Constructing a Language: A Usage-Based Theory of Language Acquisition. Harvard University Press, Cambridge
46. Turkle S (2011) Alone together: why we expect more from technology and less from each other. Basic Books, New York
47. Vergados D, Alevizos A, Mariolis A, Caragiozidis M (2008) Intelligent Services for Assisting Independent Living of Elderly People at Home. In: Proceedings of the international ACM conference on pervasive technologies related to assistive environments (PETRA'08), Athens, Greece, 79:1–4
48. Wahlster W (2003) SmartKom: Symmetric multimodality in an adaptive and reusable dialogue shell. In: Krahl R, Günther D (eds) Proceedings of the human computer interaction status conference 2003, Berlin, Germany, pp 47–62
49. Waibel A, Steusloff H, Stiefelhagen R, Watson K (2009) Computers in the human interaction loop. In: Waibel A, Stiefelhagen R (eds) Computers in the human interaction loop. Springer, London, pp 3–6
50. Wendemuth A, Biundo S (2012) A companion technology for cognitive technical systems. In: Esposito A, Esposito AM, Vinciarelli A, Hoffmann R, Müller VC (eds) Cognitive behavioural systems, LNCS 7403. Springer, Berlin, pp 89–103
51. Wendemuth A, Braun J, Michaelis B, Ohl F, Rösner D, Scheich H, Warnemünde R (2008) Neurobiologically inspired, multimodal intention recognition for technical communication systems (NIMITEK). In: Proceedings of the 4th IEEE tutorial and research workshop on perception and interactive technologies for speech-based systems, PIT 2008, Kloster Irsee, Germany, pp 141–144
52. Wilks Y (2007) Is there progress on talking sensibly to machines? Science 318(5852):927–928
53. Wilks Y (2010) Introducing artificial companions. In: Wilks Y (ed) Close engagements with artificial companions. Key social, psychological, ethical and design issues. John Benjamins: Amsterdam, pp 11–20
54. Winkler I, Denham SL, Nelken I (2009) Modeling the auditory scene: predictive regularity representations and perceptual objects. Trends Cognit Sci 13(12):532–540
55. Ziemke T (2008) On the role of emotion in biological and robotic autonomy. BioSystems 91:401–408

Chapter 5
Communication Sequences and Survival Analysis

Carl Vogel

Abstract Two new methods of analyzing dialogue interactions are outlined. One method depends on abstract representations of dialogue events as symbols in a formal language. This method invites analysis of the expressivity requirements of dialogue grammar, as well as distribution analysis of dialogue event symbol sequences. The method is presented in relation to a temporal construction from regular languages, one which supports increasingly fine granularity of temporal analysis. The other method proposed is also temporally oriented. It also depends on dialogue events and dialogue states, and proposes to analyze causal relations among dialogue events through survival analysis. These methods are suggested as additions to the extant repertoire of approaches to understanding the structure and temporal flow of natural dialogue. Additional methods of analysis of natural dialogue may contribute to deeper understanding of the phenomena. With deeper understanting of natural dialogue one may hope to more fully inform the construction of believable artificial systems that are intended to engage in dialogue with a manner close to human interaction in dialogue.

5.1 Introduction

Artificial agents with socially believable communication strategies are most naturally informed by human communication behaviors. Recent research into patterns of human communication has examined quantification of engagement. Degree of repetition is a useful proxy measure of engagement, and within a number of studies, comparison of levels of self-repetition and allo-repetition between turns in actual dialogue and randomized counterparts of dialogue provides a means to state when

C. Vogel (✉)
Centre for Computing and Language Studies, School of Computer Science
and Statistics, O'Reilly Institute, Trinity College Dublin, The University of Dublin,
Dublin 2, Ireland
e-mail: vogel@tcd.ie

© Springer International Publishing Switzerland 2016
A. Esposito and L.C. Jain (eds.), *Toward Robotic Socially Believable
Behaving Systems - Volume II*, Intelligent Systems Reference Library 106,
DOI 10.1007/978-3-319-31053-4_5

47

repetition in dialogue exceeds that which one might expect in random baselines [15, 19, 20]. Such repetition effects have been shown to correlate with task-oriented success as proxy measures of communication success in dialogue [16, 17]. This article examines idealizations alternative specifications of baselines for believable social interaction. Two families of approaches are considered: firstly, sequence analysis, drawing on formal language theoretic idealizations of dialogue; secondly, survival analysis. Both of these directions embrace the integrative approach to believable artificial communication strategies recently argued for in the literature [7].

5.2   Dialogue Symbol Sequences

It is possible to analyze dialogue interactions via idealizations of the sort of content that is expressed at any particular moment. This may be at the rich level of description afforded by conversational analysis [18] or at more abstract levels [2, 13]. For example, the temporal flow of patterns of silence and non-silence are useful in explicating the structure of discussions and their social dynamics [3]. Consider the regular grammar in (5.1).[1] The terminal symbol s^i represents a silence of speaker i,[2] and the terminal symbol v^i represents a vocalization of speaker i. The set of productions is given in (5.2)–(5.8). The superscripts n and m on the nonterminal and terminal symbols vary over the set $\{1, 2\}$, for two speakers; thus the production rules with these variable superscripts are schema that abbreviate the enumeration of all of the (finite number of) possible instantiations of those variables.

$$\langle S, \{P^1, P^2\}, \{s^1, s^2, v^1, v^2\}, \{2, 3, 4, 5, 6, 7, 8\}\rangle \tag{5.1}$$

$$P^n \rightarrow v^n \tag{5.2}$$

$$P^n \rightarrow s^n \tag{5.3}$$

$$P^n \rightarrow v^n P^m \tag{5.4}$$

$$P^n \rightarrow s^n P^m \tag{5.5}$$

$$S \rightarrow P^1 \tag{5.6}$$

$$S \rightarrow P^2 \tag{5.7}$$

$$P^n \rightarrow S \tag{5.8}$$

The language generated is as in (5.9).

$$\{w | w \in (s^1 + s^2 + v^1 + v^2)^+\} \tag{5.9}$$

---

[1]In this structure, the first element, S, is a special non-terminal symbol, the start symbol; the next element the remaining non-terminal symbols; the third element is a set of terminal symbols; the final element is a set of productions, here pointers to the productions. Each set in this structure is finite. The language generated is infinite.

[2]The representation s^i is adopted as a positive representation of silence, thinking of this as distinct from the usual representation of the empty string (ε). Shortly, representation of distinct sorts of silence will be introduced.

Notice that this is exactly Σ^+. This is trivial in the sense that any non-empty sequence of terminal symbols in the alphabet of the grammar is licensed as well-formed. However, this is a reasonable model at this level of granularity of description of dialogue contributions—the extent to which the assertion that any sequence of vocalizations and silences is well formed is wrong is the extent to which one pays attention to subcategories of vocalization and silence [18].

Nonetheless, this structure is not satisfying. Although it does capture contributions of the two speakers (and generalizes with additional symbols to any number of speakers), it does not capture the simultaneity of contributions of each. It is adequate to think of discrete moments of time (generalizing from finite state approaches to temporality and the semantics of aspect [8]). Define a grammar as in (5.10) for the sequence of vocalizations and silences of each dialogue participant, i.

$$\langle S, \{P^i\}, \{v^i, s^i\}, \{11, 12, 13, 14, 15\}\rangle \tag{5.10}$$

$$P^i \rightarrow v^i \tag{5.11}$$

$$P^i \rightarrow s^i \tag{5.12}$$

$$P^i \rightarrow v^i P^i \tag{5.13}$$

$$P^i \rightarrow s^i P^i \tag{5.14}$$

$$S \rightarrow P^i \tag{5.15}$$

The language generated is given in (5.16).

$$L^i = \{w | w \in (s^i + v^i)^+\} \tag{5.16}$$

Given L^m and L^n, instances of (5.10), define the *superposition* (&) of L^m and L^n ($L^m \& L^n$) as in (5.17) [8].

$$L^m \& L^n = \bigcup_{k \geq 1} \{(\alpha_1^m \cup \alpha_1^n) \dots (\alpha_k^m \cup \alpha_k^n) | \alpha_1^m \dots \alpha_k^m \in L^m, \alpha_1^n \dots \alpha_k^n \in L^n\} \tag{5.17}$$

This depends on a notion of component sequences, for example, as in (5.18), where each component is a set of symbols. Components may stand as string symbols in their own right. The grammar that generates this can be obtained from the grammar (5.10) by embedding each terminal symbol in the production rules inside a set.

$$L^i = \{w | w \in (\{s^i\} + \{v^i\})^+\} \tag{5.18}$$

In turn, this permits a definition of componentwise union ($\cup$) of strings of equal length (k), where Σ_i is the terminal vocabulary of L^i, and $\aleph^i \subseteq \{\beta | \beta \in \Sigma_i\}$, and $\sigma^i = \langle \aleph_1^i, \dots, \aleph_k^i \rangle$ as in (5.19).

$$\sigma^m \cup \sigma^n = \langle \aleph_1^m \cup \aleph_1^n, \dots, \aleph_k^m \cup \aleph_k^n \rangle \tag{5.19}$$

Each component, a composite symbol, of a string σ represents a moment of time, and the componentwise union of strings corresponds to the contribution of each participant at the same moment in time. Thus, given the sequence of contributions (5.20) and (5.21), the componentwise union involves two moments in which the speakers alternate silence and speaking, followed by one moment of overlapping vocalization (5.22).

$$\sigma^1 = \langle\{v^1\}, \{s^1\}, \{v^1\}\rangle \in L^1 \tag{5.20}$$

$$\sigma^2 = \langle\{s^2\}, \{v^2\}, \{v^2\}\rangle \in L^2 \tag{5.21}$$

$$\sigma^1 \cup \sigma^2 = \langle\{v^1, s^2\}, \{s^1, v^2\}, \{v^1, v^2\}\rangle \tag{5.22}$$

On this model, all symbols within a component are understood as simultaneous; each component of a string occupies the same duration as each other unit in the string. At this point, it is useful to note that there are at least two ways in which this representation may be understood. One is that the symbol v^i stands for the vocalization *qua* vocalization of the utterance (linguistic or otherwise) that it names, wherein its duration is the composite duration of the duration of each constituent vocalization, and each constituent vocalization is a distinct proper part of v^i. Another interpretation is that v^i is the proposition that a total vocalization of some sort is happening.[3] On this interpretation, because the proposition is stative, each proper sub-interval of the duration of v^i is also an interval in which the total vocalization is happening, and v^i holds true, even if the total vocalization is incomplete over the sub-interval. This property does not hold as cleanly for the first interpretation, as at a proper sub-interval of the duration of v^i, v^i itself cannot be correctly said to have been vocalized; rather, a proper part of v^i is vocalized over the sub-interval. That is, a vocalization of "Sam has baked a loaf of bread" occupies some total duration, t. However, a proper sub-interval of that duration is not that of a vocalization of "Sam has baked a loaf of bread", but perhaps of something like "Sam has". On the other hand, the proposition that "Sam has baked a loaf of bread" is being uttered holds true during both $\Delta = \delta$("Sam has baked a loaf of bread") and $\Delta' = \delta$("Sam has"), where Δ' is a sub-interval of Δ.[4] The same consideration applies to the silence symbol (or additional symbols for dialogue events).

In order to accommodate varying temporal granularity of strings it is necessary to more clearly associate temporal information with strings. This may be done by associating a clock, C, with the grammar, as in (5.23).

$$\langle S, N, T, P, C \rangle \tag{5.23}$$

---

[3]On the first interpretation, v^k may represent an utterance such as, "the cat is on the mat." On the second interpretation, v^k may represent the proposition, "'the cat is on the mat' is being uttered". In the first case, v^k stands for the vocalization as a vocalization, and in the second case it stands for the proposition that the vocalization is under way.

[4]During $\Delta^\dagger = \delta$("Sam has"), $\Delta^\dagger$ a proper sub-interval of $\Delta^\ddagger = \delta$("Sam has water"), the proposition that a vocalization of "Sam has baked a loaf of bread" is happening is not true.

The clock is used to determine a start time and a constant "frame rate" for a grammar, using the motion picture metaphor that has been used in finite state temporality [8]. Using this analogy, a string corresponds to a film clip, each terminal symbol amounting to the observations recorded within a corresponding frame of the clip. Each string has a start time, which is at whatever offset from the clock's zero, and each component (each set that contains a terminal symbol), has the same duration, since the clock is assumed to set the start time and frame rate. This model makes it possible to devise a new sort of component union, one that depends not on strings having the same number of components, as in (5.19), but rather on strings having the same starting time and ending time. Two strings with the same start time and end time and which also share the same frame rate will have the same number of components, and their temporal component union will be the same as with (5.19), except for the additional temporal information.

Consider the three putative strings depicted in (5.24), and assume that each of σ^1, σ^2 and σ^3 share temporal endpoints. Firstly, notice that the putative string σ^3 is ruled out as a possible string, since it is depicted as having variable frame durations, contrary to the assumptions described above. In contrast, σ^1 and σ^2 are shown as having constant frame rates within them.

$$\sigma^1 = \langle \{v^1\}\{s^1\}\{v^1\}\{v^1\}\{s^1\}\{v^1\}\rangle$$
$$\sigma^2 = \langle \{\ v^2\ \}\{\ v^2\ \}\{\ v^2\ \}\rangle \qquad (5.24)$$
$$\sigma^3 = \langle \{\ s^3\ \}\{\ v^3\ \}\{s^3\}\{\ s^3\ \}\rangle$$

Now, the question is what a temporal componentwise union of σ^1 and σ^2 can be. Since the terminal symbols are interpreted as propositions in the way described above, a proposition that holds of a duration also holds during sub-intervals of that duration. On the alternative interpretation of the symbols described above, this construction would not be accurate in the start and stop times of symbols distributed across shorter constituent durations.

$$\sigma^1 \overset{t}{\cup} \sigma^2 = \langle \{v^2, v^1\}\{v^2, s^1\}\{v^2, v^1\}\{v^2, v^1\}\{v^2, s^1\}\{v^2, v^1\}\rangle \qquad (5.25)$$

A specification of temporal compontentwise union is given below (5.26) for the basic case in which σ^m and σ^n are co-terminus, but σ^n consists of a single component ($\sigma^n = \langle \aleph^n\rangle$). In this case, there is a component for each component of σ^m, and the constituent durations are determined by the components of σ^m.

$$\sigma^m \overset{t}{\cup} \langle \aleph^n\rangle = \langle \aleph^m_1 \cup \aleph^n, \ldots, \aleph^m_k \cup \aleph^n\rangle \qquad (5.26)$$

Let δ be a measure of the duration of a symbol, $\alpha \in \Sigma$, or of a component χ in a string, χ^n, of length n. As specified above, the clock applies to all strings of a language (5.27).

$$L_\Delta = \{w | w \in (\{\})^*, \delta(\{\}) = \Delta\} \qquad (5.27)$$

If $\sigma = \Sigma^n$, then $\delta(\sigma) = n \cdot \delta(\alpha)$. The temporal componentwise union allows strings from languages with distinct frame rates to be joined. The distinction between vocalization and silence may be further articulated (e.g. listening silence, ignoring silence, etc.), but also other dimensions may be addressed. The intuition is that there is a language of this sort for each dimension of communication in which each agent may engage (e.g. lexical, hand-gesture, facial expression, intonation, laughter, coughs, etc.). Each of those dimensions for each agent brings its own grammar [4, 14]. For each of these dimensions, a separate grammar introduces terminal symbols for the sorts of vocalizations or other form of expression sanctioned in that dimension. For example, the grammar of (5.28) introduces terminal symbols for (largely involuntary) bodily functions that accompany communication: a for sneezes, b for burps, c for coughs.[5] Of course, these same functions may be dissembled as involuntary, in which case they have a deeper communicative function. However, even in the case of involuntary actions of these sorts, there is a frequent and natural tendency to try to suppress these to "least worst" (if not "most appropriate") moments of interaction for them to happen. As such, they convey information about their agent's evaluation of the unfolding interaction.

$$\langle S, \{P^i\}, \{v^i, s^i\}, \{29, 30, 31, 32, 33, 34, 35, \}, C\rangle \tag{5.28}$$

$$P^i \to \{a^i\} \tag{5.29}$$

$$P^i \to \{b^i\} \tag{5.30}$$

$$P^i \to \{c^i\} \tag{5.31}$$

$$P^i \to \{a^i\}P^i \tag{5.32}$$

$$P^i \to \{b^i\}P^i \tag{5.33}$$

$$P^i \to \{c^i\}P^i \tag{5.34}$$

$$S \to P^i \tag{5.35}$$

One of the goals of this mode of analysis is to support reasoning about the nature, including computational complexity, of dialogue interactions. In modeling dialogue interactions with regular grammars,[6] the least expressive computational framework for capturing infinite sequence possibilities is invoked. Where Σ is the overall set of terminal symbols from the superposition of the grammars for each dimension, this does not entail the assertion that the set of interactions is best described by Σ^* or

---

[5]The symbol a is meant to be suggestive of the English onomatopoeic expression, "achoo".

[6]I have not proven that the addition of clocks as proposed here does not increase the expressivity of the framework beyond the expressivity of regular languages. The intuition behind the argument that regular grammars with clocks of this sort remain regular is that the "grouping" of symbols within sets, in an initial grammar, only ever includes a single terminal symbol, never a non-terminal symbol; thus, the effect of bracket matching on either side of a recursive use of a non-terminal symbol (such as is the prototype of a context-free grammar rule) does not occur. The information encoded by the clock for a language may be captured in a constructed regular language without a clock by adding a terminal symbol for the starting time, and another symbol for subsequent "ticks" of the clock. Also, recall that regular languages are closed under both intersection and union [12].

Σ^+, that any sequence of moves at all is well formed (although this may be the case, and is not ruled out, a priori. Rather, the potential for encoding that some sequences are possible and others are not (as has been suggested [18]) is available up to the level of expressivity of regular grammars. It has been argued that some fragments of natural dialogue interactions are inherently more expressive than regular, and beyond context-free [13]. However, it is not fully clear that the extant arguments are not equivalent to the fallacies of putative expressivity that have been pointed out in other areas of natural language [10]. If regular expressivity is adequate, then, given the propositional interpretation of the terminal symbols used here, then the current framework has the same advantage open to finite state temporality in modeling entailment with the language inclusion problem, whether one set of sequences of propositions is a subset of another set of sequences of propositions, [9], since inclusion is decidable for regular languages [12]. Nonetheless, even if the abstract language of dialogue interactions is provably of greater than regular expressivity,[7] it is useful to think about dialogue interactions in these terms. For example, recent work has explored distributions of symbol sequences effectively involving temporal componentwise unions of social and linguistic signals, in order to study the discourse marking effects of social signals [2]: it was found that sequences inclusive of a topic change symbol significantly more frequently also contain overlap symbols than symbols representing long pauses.[8] Languages specified in the way described here enumerate sequences that represent possible dialogues in which all sequences are as likely as every other sequence. A point, however, is that one may conduct distributional analysis of the grammatical sequences [2] in relation to the baseline model of a uniform distribution. It is then possible to articulate more fine-grained relations about the proximity of some symbols to other symbols in the categories of sequences that dominate the actual distributions. This is orthogonal to being able to study the difference between allowable sequences and ill-formed sequences, towards understanding the expressive requirements of dialogue grammar.

5.3   Survival Analysis

The preceding discussion attached starting points and frame rates to strings of components of symbols, each symbol interpreted as the proposition that a dialogue move of some sort or other has happened in the dialogue modelled. It has been noted that it is of interest to examine these sequences in order to learn about relationships they embed—for example, the greater proximity of some categories of symbols to other reference point symbols (such as topic change). In the mode of operation described above the frame rates are taken to be constant throughout a string, but through temporal componentwise union, one may obtain strings with a "faster" frame rate, a finer level of granularity of temporal analysis. Study of dialogue phenomena also

---

[7]At greater levels of expressivity, context-free and higher, the inclusion problem is not decidable [12].

[8]In the present work, an overlap is equivalent to a component like $\{v^1, v^2\}$.

benefits from scrutiny without the assumption of fixed frame rates. Analysis drawn from survival analysis may be relevant, and again it is necessary to establish relevant baselines so that actual data may be compared to the baselines.

The assertion that survival analysis may be appropriate to the study of dialogue is drawn from the observation that if an interaction qualifies as a well formed dialogue, then one can be as certain of some facts as of death and taxation: utterances will happen; silences will happen; gestures will happen; overlaps might happen; laughter might happen; coughs might happen. Not all events are certain, but some are. However, even the propositions related to uncertain events are temporally bounded—an overlap might happen, but if it does, it will eventually end. This suggests an alternative way of exploring the relationships among dialogue events via timed symbol sequences that represent the containing dialogues. In the simplest version of the idea, one examines the "survival" of symbols of interest in relation to alternative treatments: other symbols that precede "death". That is, topics will happen during a dialogue and will inevitably terminate within the dialogue. The question is whether survival analysis may be used to quantify which other symbols' appearance in immediate or relatively prior positions in the sequences that include topic termination are more or less likely to hasten, correlate with or otherwise predict topic termination.

5.3.1 General Method

To engage in this sort of analysis a coding of data may be conducted using standard statistical tools for analysis of "time to events". Again, actual data and effects may be compared with effects that obtain in data constructed using random generators. Suppose that one is interested in whether the occurrence of laughter correlates with topic change, then it is possible to measure the survival of topics in relation to whether laughter occurs within some interval from the end of a topic. For each topic the duration of the topic is recorded. Using the same starting point measure as for the duration of the topic, the duration of time without laughter may be recorded. For some d either that varies with the topic length (e.g. half the duration of the duration of the topic) or a constant (e.g. 15 s), a factor related to laughter being present close enough to the topic termination to be related to it may be created (5.36).[9] If d is equal to the topic duration, then LM is a binary indication of whether laughter occurs within the topic or not.

$$LM = ((T - \neg L) > 0) \wedge ((T - \neg L) \leq T \cdot d) \tag{5.36}$$

---

[9]In order to subtract away a constant c from T using (5.36), it is sufficient to specify d as $\frac{T-c}{T}$; to test whether $T - \neg L$ is less than some constant c, it suffices to specify d as $\frac{c}{T}$.

The terms of (5.36) are: LM, a binary indication of whether laughter matters for the topic[10]; T, the time from the onset to the termination of the topic; $\neg L$, the time from the onset of the topic without laughter[11]; d, described above. Similar factors may be constructed from other signals that one might wish to correlate with topic change (e.g. coughing, overlap, etc.). One may then apply LM and its counterparts for the other possible dialog events in an analysis of the "survival" of topics with and without laughter, for example, close enough to the topic termination (for whatever measure of d that is appropriate) to plausibly be causally related.[12] The interest is in whether there is a difference in the survival of topics between topics with instances of laughter (or other dialogue events) commencing d-close to the topic end and topics without such laughter (or other dialogue events). Equivalently, one may study the survival of laughter conditioned upon factors derived from other dialogue events.

5.3.2 An Example

Recent work in the analysis of laughter as a discourse marker conveying information about topic change explored the likelihood of laughter in the windows of time before and after topic changes. Whether the windows were defined relative to the overall duration of the topic (for example, considering the first quarter of a topic's duration or the topic's final quarter) or defined in absolute temporal terms (e.g. the first 15 s or last 15 s of a topic), the same effects emerged during the analysis of more than one multi-modal dialogue corpus: instances of laughter were less likely at the start of a new topic than at a topic's end [1, 11].[13] This section develops an example application of the method described above to the TableTalk dataset analyzed within the recent works just mentioned [5].

In this dataset, free conversation among five participants over three sessions are recorded. The conversation setting was informal, within the Advanced Telecommunication Research Labs in Japan. The participants included three women (Australian, Finnish and Japanese) and two men (Belgian and British). Topic annotations (116 distinct topics) and annotations of laughter (719 instances) are provided with the corpus, which spans 3 hours and 30 min of chat.

Initially, consider $\neg L$ to be the time within a topic before the onset of its last moment of laughter, and take d to be 0.10, so that $T \cdot d$ (in (5.36)) refers to 10 %

---

[10]That is, this is true if laughter is close enough to the change of topic to be reasonably hypothesized as a discourse marker.

[11]Non-laughter, which terminates with the onset of laughter, is constructed as a dialogue event, so that one may examine the survival rate of non-laughter.

[12]Interestingly, given that dialogue participants may be imputed to all have a sense of when discussion of a topic is coming to an end, if this sensation does give release to laughter, or other signals, that correlate with topic end, then there is a case to be made for temporally inverted causation: the topic end may cause the laughter which precedes it.

[13]Laughter, it turns out, is not a precise signal of topic ending, but if laughter is present, it serves as a very good signal that the topic of discussion has not just started.

of the topic's duration: where LM is true, the topic is one in which the last moment of laughter occurs in the final 10% of the topic's duration. The plot in the in the left-hand side of Fig. 5.1 depicts the time (in seconds) to the end of topic for topics where the last laugh occurs in the final 10% of its duration in comparison with the topic durations without laughter so placed: topics for which the laughter occurs in the final 10% are significantly longer in duration than the complement.[14] This extends to contrasts involving the existence of final laughter in shorter durations relative to the topic duration as well: the final 5% ($p < 0.005$), final 2.5% ($p < 0.005$), final 1.25% ($p < 0.005$). The direction switches for absolute duration measurements. The plot in the right-hand side of Fig. 5.1 compares topic durations for topics that have final laughter in the last ten seconds with topic durations for the complement. The former are significantly ($p < 0.05$) shorter than the latter. The comparisons related to absolute durations slightly less than 10 s and slightly more than 10 s exhibit the same trend, but without statistical significance. A fact common to both of the relative and absolute examinations is that ten topics did not contain laughter, and as a group those topics were significantly shorter in duration than topics that contained laughter. However, focusing analysis on just the topics containing laughter does not disrupt the overall pattern of results—only the test of topic durations with final laughter in the final 5% loses significance, but the same direction and approximate strength of effect holds in each other case.

There is a significant ($P < 0.0001$) positive correlation (0.51) between laughter counts and topic duration, but this correlation is weaker, yet still strong (0.48), and still significant ($P < 0.0001$) among only topics containing laughter. Thus, the analysis of laughter in the final d percentages of topic durations may be revealing little more than that topics containing more laughter in the final d percent contain more laughter overall. Examining topics that have laughter in the final d percent ($d < 10$, for the percentages indicated above), or in the final d seconds, there are significantly more instances of laughter overall for the topic than for topics that do not have laughter in those topic-final moments (Wilcox test, $p < 0.005$). Thus, the greater presence of laughter overall during the topics may explain the plot on the left in Fig. 5.1, but not the data plotted on the right, where it is shown that topics with laughter in the final ten seconds are shorter in duration than the complement.

5.3.3 Discussion

The differences noted above in the time to the end of the topics for topics that have instances of laughter in terminating sequences defined in absolute terms in relation to those with terminating sequences defined relative to the overall topic length is

---

[14]Using the one-tailed asymptotic log-rank test $p < 0.005$. See `surv_test` within the `coin` package within R (http://cran.r-project.org/web/packages/coin/index.html—last verified, July 2015) created by Torsten Hothorn, Kurt Hornik, Mark A. van de Wiel and Achim Zeileis.

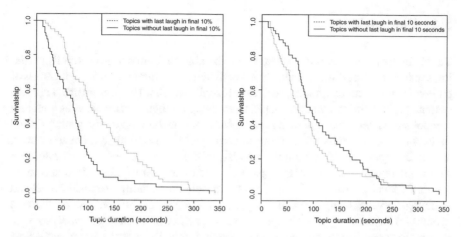

Fig. 5.1   Time in seconds to end of topic for topics with and without terminating laughter

intriguing. For some types of events, their causal impact is most likely to be evident within a temporal window defined in absolute terms (a falling is likely to have been caused by a tripping within a few seconds prior to the falling rather than by a tripping within the final few percent of the walk). For others, such causal relations are more evident in relational definitions (fatal lung cancer is more likely to have been caused by having taken up intensely smoking cigars in the final ten percent of a life than in the final three years of life without reference to the total life span). The results noted above show that the time to the end of topics is extended by laughter (the plot on the left in Fig. 5.1) but also suggest that the presence of laughter creates a window in which it is also relatively easy for topic to change sooner, hastening time to the end of the topic (the plot on the right in Fig. 5.1).[15] These results are far from conclusive, but are suggestive of additional questions to explore in isolating the categories of laughter that hasten or delay topic termination – for example, evenness versus clumpiness in the distribution of laughter across a topic's duration.[16] Of course, by construction, the grouping of topics here does not correspond to treatments as typically analyzed within survival analyses. In fact, it is a purpose of this work to demonstrate that survival analysis may be fruitfully adapted to the purposes of analyzing patterns within sequences of dialogue moves.

---

[15]It is important to remember that the topic durations in the two representative plots relate to the same topics, but under distinct classifications.

[16]It is also necessary to apply this sort of analysis to additional dialogue corpora in order to understand the robustness of the effects found for the TableTalk corpus.

5.4 Conclusions

That believable autonomous systems must be able to demonstrate naturalistic dialogue interactions follows from the fact that dialogue interactions inform diagnoses of pathological human conditions, such as schizophrenia [6]. The present article is programmatic in defining two approaches to studying dialogue interactions as temporal symbol sequences. One approach is anchored in formal language theory analysis and focuses on the generative framework that can yield appropriate symbol sequences, as well as supporting distributional analysis of the sequence sets without reference to grammars directly. The other approach handles temporal information at a more fine-grain level of detail and considers causal relations among propositions related to dialogue events using survival analysis, given temporal distributions of dialogue symbols. The motivation for proposing these paradigms of dialogue analysis is to pursue them as a way of gaining greater insight into the structure and temporal flow of dialogue events. If the nuances of dialogue structure and temporal flow can be better understood for human dialogues from a wide range of genres then there is increased possibility for incorporating this knowledge in automated systems that are intended to have believable interaction possibilities.

Acknowledgments I am grateful to Francesca Bonin, Nick Campbell, Anna Esposito and Emer Gilmartin. They have no responsibility for any wrong-headedness that appears here, though. This research is enhanced by supported from Science Foundation Ireland through the CNGL Programme (Grant 12/CE/I2267) in the ADAPT Centre (www.adaptcentre.ie) at Trinity College Dublin. The ADAPT Centre for Digital Content Technology is funded under the SFI Research Centres Programme (Grant 13/RC/2106) and is co-funded under the European Regional Development Fund.

References

1. Bonin F, Campbell N, Vogel C (2014) Time for laughter. Knowl Based Syst 71:15–24
2. Bonin F, Vogel C, Campbell N (2014) Social sequence analysis: temporal sequences in interactional conversations. In: Péter B (ed) Proceedings of the 5th IEEE international conference on cognitive infocommunications, pp 403–406
3. Bouamrane MM, Luz S (2007) An analytical evaluation of search by content and interaction patterns on multimodal meeting records. Multimed Syst 13:89–102. doi:10.1007/s00530-007-0087-8
4. Bunt H, Alexandersson J, Chae JW, Fang AC, Hasida K, Petukhova O, Popescu-Belis A, Traum D (2012) ISO 24617-2: a semantically-based standard for dialogue act annotation. In: Proceedings of the LREC
5. Campbell N (2009) An audio-visual approach to measuring discourse synchrony in multimodal conversation data. In: Proceedings of the interspeech
6. Covington MA, He C, Brown C, Naçi L, McClain JT, Fjordbak BS, Semple J, Brown J (2005) Schizophrenia and the structure of language: the linguists view. Schizophr Res 77:85–98
7. Esposito A, Esposito AM, Vogel C (2015) Needs and challenges in human computer interaction for processing social emotional information. Pattern Recognit Lett (to appear). http://dx.doi.org/10.1016/j.patrec.2015.02.013
8. Fernando T (2004) A finite-state approach to events in natural language semantics. J Logic Comput 14(1):79–92. doi:10.1093/logcom/14.1.79

 9. Fernando T (2007) Observing events and situations in time. Linguist Philos 30(5):527–550. doi:10.1007/s10988-008-9026-1
10. Gazdar G, Pullum G (1985) Computationally relevant properties of natural languages and their grammars. Technical Report CSLI-85-24, stanford: center for the study of language and information
11. Gilmartin E, Bonin F, Vogel C, Campbell N (2013) Laugher and topic transition in multi-party conversation. In: Fernández R, Isard A (eds) Proceedings of the SIGDIAL 2013 conference. Association for computational linguistics, Metz, pp 304–308. http://www.aclweb.org/anthology/W/W13/W13-4045
12. Hopcroft JE, Ullman JD (1979) Introduction to automata theory, languages, and computation. Addison-Wesley Publishing Co., Reading
13. Păun G (1984) Modelling the dialogue by means of formal language theory. In: Vaina L, Hintikka J (eds) Cognitive constraints on communication: representations and processes. Reidel, Dordrecht, pp 363–372
14. Petukhova V, Bunt H (2009) Towards a multidimensional semantics of discourse markers in spoken dialogue. In: Proceedings of the eighth international conference on computational semantics. Association for computational linguistics, Stroudsburg, IWCS-8 09, pp 157–168
15. Ramseyer F, Tschacher W (2010) Nonverbal synchrony or random coincidence? How to tell the difference. In: Esposito A, Campbell N, Vogel C, Hussain A, Nijholt A (eds) Development of multimodal interfaces: active listening and synchrony. LNCS. Springer, Berlin, pp 182–196
16. Reitter D, Moore J (2007) Predicting success in dialogue. In: Proceedings of the 45th annual meeting of the association of computational linguistics. Association for computational linguistics, pp 808–815
17. Reitter D, Keller F, Moore J (2006) Computational modeling of structural priming in dialogue. In: Proceedings of the human language technology conference of the North American chapter of the ACL. Association for computational linguistics, pp 121–124
18. Schegloff EA (2007) Sequence organization in interaction: a primer in conversational analysis, vol 1. Cambridge University Press, Cambridge
19. Vogel C (2013) Attribution of mutual understanding. J Law Policy 21(2):377–420
20. Vogel C, Behan L (2012) Measuring synchrony in dialog transcripts. In: Esposito A, Esposito AM, Vinciarelli A, Hoffmann R, Müller VC (eds) Cognitive behavioural systems, vol 7403. LNCS. Springer, Heidelberg, pp 73–88

Chapter 6
The Relevance of Context and Experience for the Operation of Historical Sound Change

Jonathan Harrington, Felicitas Kleber, Ulrich Reubold and Mary Stevens

Abstract This paper is concerned with explaining how historical sound change can emerge as a consequence of the association between continuous, dynamic speech signals and phonological categories. The relevance of this research to developing socially believable speech processing machines is that sound change is both cognitive and social and also because it provides a unique insight into how the categories of speech and language and dynamic speech signals are inter-connected. A challenge is to understand how unstable conditions that can lead to sound change are connected with the more typical stable conditions in which sound change is minimal. In many phonetic models of sound change, stability and instability come about because listeners typically parse—very occasionally misparse—overlapping articulatory movements in a way that is consistent with their production. Experience-based models give greater emphasis to how interacting individuals can bring about sound change at the population level. Stability in these models is achieved through reinforcing in speech production the centroid of a probability distribution of perceived episodes that give rise to a phonological category; instability and change can be brought about under various conditions that cause different category distributions to shift incrementally and to come into contact with each other. Beyond these issues, the natural tendency to imitation in speech communication may further incrementally contribute to sound change both over adults' lifespan and in the blending of sounds that can arise through dialect contact. The general conclusion is that the instabilities that give rise to sound change are an inevitable consequence of the same mechanisms

J. Harrington (✉) · F. Kleber · U. Reubold · M. Stevens
Institute of Phonetics and Speech Processing, Ludwig-Maximilians-University
of Munich, Munich, Germany
e-mail: jmh@phonetik.uni-muenchen.de

F. Kleber
e-mail: kleber@phonetik.uni-muenchen.de

U. Reubold
e-mail: mes@phonetik.uni-muenchen.de

M. Stevens
e-mail: reubold@phonetik.uni-muenchen.de

© Springer International Publishing Switzerland 2016
A. Esposito and L.C. Jain (eds.), *Toward Robotic Socially Believable Behaving Systems - Volume II*, Intelligent Systems Reference Library 106, DOI 10.1007/978-3-319-31053-4_6

that are deployed in maintaining the stability between phonological categories and their association with the speech signal.

A fundamental challenge in phonetics and the speech sciences is to understand how the interleaved movements of continuous speech signals are associated with categories such as consonants and vowels that function to distinguish between word meanings. The dichotomy between these two levels of representation comes about because on the one hand any speech utterance is highly context-dependent but on the other hand languages distinguish words by means of a finite cipher of abstract phonological units that can be permuted in different ways. There is abundant evidence for the context-dependent nature of speech. The same utterance can vary dramatically depending on the speaking situation and environment—whether talking to friends or in a more formal speaking situation, whether there is background noise or quiet [92]. Speech is also context-dependent because speech sounds are synchronised in a temporally overlapping way: producing speech is a shingled movement [144], so that any particular time slice of the speech signal provides the listener with information about speech sounds that have been, and that are about to be produced and in a way that is also different depending on prosodic factors to do with syllable position and the stress or prominence with which syllables are produced [9]. These are then many of the reasons why temporally reversing a speech signal of *stack* does not lead to an unambiguous percept of *cats*. Moreover, the context-dependence is not just a function of the sluggishness of the vocal organs in relation to the speed with which speech is produced, but also communicates much about the social and regional affiliations of the speaker [60, 96]. At the same time, phonological abstraction from these details is fundamental to human speech processing: there is a sense in which the different words *stack*, *cats*, *acts*, and *scat* are permutations of the same four phonological units or phonemes. There is now extensive evidence that children learn both levels of representation in speech communication: they are on the one hand responsive to acoustic information arising from continuous movement in the speech signal. But simultaneously they acquire the ability to perform phonological abstraction which allows them to recombine abstract phonological units to produce words that they have not yet encountered (e.g. Beckman et al. [10]).

The task in this paper is to make use of the existing knowledge about the connections between these two very different ways of representing speech in order to explain the operation of sound change; and in turn to use what is known about sound change to place constraints on the type of architecture that is possible and required for linking these physical and abstract levels of speech sound representation. The focus will be on what the Neogrammarians of the 19th century [115, 118] termed regular sound change which they considered to be gradual and imperceptible and to apply to all words; this type of sound change was for them distinct from analogical change which was irregular, phonetically abrupt (in the sense that the change was immediate, not gradual), lacked phonetic motivation, and often applied to only a handful of words (see e.g. Hualde [68] for some examples of change by analogy).

Modelling sound change is relevant to understanding how cognitive and social aspects of human speech processing are connected. The association between these two domains has been largely neglected in the 20th century, partly because whereas generative theories of phonology draw upon highly idealised data (typically pho-netic transcriptions) to develop a grammar consisting of re-write rules based on supposed linguistic universals and operating on a lexicon containing the minimum of information to represent the distinctions between words, sociolinguistic models by definition are concerned with explaining how variation in speech is conditioned by factors such as gender, age, and social class, factors that are beyond the scope of generative models. In the last 10–15 years, there have, however, been increas-ing attempts to reconcile these two positions largely within so-called usage-based, episodic or exemplar models of speech (see e.g. Docherty and Foulkes [31] for a recent review) that are derived from models of perception in cognitive psychology (e.g. Hintzman [65]) and that give much greater emphasis to the role of memory in human speech processing. Exemplar theory has led to computational models of how phonological categories, the lexicon, memory and speech are inter-connected (e.g. Wedel [157]); and more generally, there has been greater emphasis in the last two decades in determining how speaker-specific attributes shape and influence cognitive aspects of human speech processing both in adults [122] and in first language acqui-sition (e.g. Beckman et al. [10], Munson et al. [103]). Understanding how social and cognitive aspects are related in human speech processing is in turn a pre-requisite for developing socially believable systems of machine speech and language processing.

The relevance of sound change in this regard is that it is evidently both cogni-tive and social. The cognitive aspects are largely concerned with the mechanisms by which phonological categories and speech signals are associated and how this asso-ciation can sometimes become unstable providing the conditions for sound change to take place. The social aspects are more concerned with how differences between speakers in their knowledge and use of language can cause sound change to spread throughout the community. These cognitive and social bases of sound change have in the last 40–50 years been pursued within largely separate frameworks concerned on the one hand with the conditions that can give rise to sound change (in particular Ohala [112]) and those that lead to its spread across different speakers on the other (in particular Labov [85, 87]). The challenge lies in developing an integrated model of sound change that draws upon insights from both approaches. This is turn can provide fresh insights into understanding how social and cognitive aspects must be inter-connected in both human and (therefore also) machine speech processing.

6.1   The Phonetic Basis of Sound Change

Much consideration has been given to the question of whether there are factors intrin-sic to the structure of the language that can bring about sound change. Such questions are typically focussed on whether there are sounds and in particular sequences of sounds that are inherently unstable, either for reasons to do with speech production or because they tend to be ineffectively communicated to the listener. Such biases in either the production or the perception of speech that predispose sounds to change should also be reflected in the typological distribution of sounds and sound sequences in the languages of the world. Thus syllables beginning with /kn, gn/ as in German *Knabe, Gnade* are rarer in the languages of the world than those beginning with /kl, gl/ [67]; and they are also involved in sound changes by which /k, g/ are deleted as in the evolution of English words that have fossilized the earlier (16th century) pronunciations in the orthography (*knave, knight, knife, gnome, gnat etc.*) but which are now in all English varieties pronounced without the initial velar stop [97].

Research concerned with the structural conditions that give rise to sound change is founded on two further ideas. Firstly, the biases that bring about sound change are directional [113]. Thus there are many sound changes by which the front vowel /y/ has developed historically from a back vowel /o/ or /u/ (e.g. the modern German *Füße*, with /fys/ in the first syllable historically from Proto-Germanic /fotiz/; hence also the English alternation *feet, foot*) but far fewer sound changes by which /y/ has retracted with the passage of time to /u/ or /o/: that is, there is a bias towards high back vowel fronting (as opposed to high front vowel retraction) both synchronically and diachronically [55]. Similarly, Guion [48] has shown that the misidentification of /ki/ as /tʃi/ which forms the basis of sound changes known as velar palatalization in numerous languages (e.g. English *chin*, but German *Kinn*) is far more likely than the perhaps unattested sound change by which /tʃi/ evolves into /ki/. Secondly, to the extent that speakers from different languages and cultures are endowed with the same physical mechanisms for producing and perceiving speech, there should be broadly similar patterns of sound change (sometimes referred to as regular sound change) in unrelated languages.

The model of sound change developed by Ohala [111–113] over a number of decades is founded upon such principles. The basis for this model is that coarticulation in speech production—that is the way in which speech sounds overlap and influence each other in time—is in almost all cases accurately transmitted between speakers and hearers. An example of coarticulation in speech production is given in Fig. 6.1 which shows the distribution of the back vowel /ʊ/ (e.g. *musste*, 'had to') in non-fronting /p/ and fronting /t/ contexts in German. In the fronting context, the tongue dorsum for /ʊ/ is further forward in the mouth due to the influence of /t/ that has its primary constriction further forward than that of /ʊ/ resulting acoustically in a raised second formant (F2) frequency.

How does the listener deal with the type of variation shown in Fig. 6.1? Experiments over several decades [41, 93, 99] show that adult listeners of the language interpret speech production in relation to the context in which it was produced. For the present example, this implies that listeners carry out a type of perceptual transformation such that they attribute the coarticulatory fronting not to the vowel itself but to the consonantal context in which it was produced. This can be demonstrated in perception experiments by synthesising a continuum between a front and back vowel in equal steps and embedding the continuum in coarticulatory fronting and non-fronting contexts. Figure 6.2 shows the results from just such an experiment for the fronting context /jist-just/, *yeast-used* (past tense) and for the non-fronting context /swip-swup/, *sweep-swoop*. (The former is a fronting context because of the /j/ and the latter a non-fronting context because the tongue dorsum for /w/ is retracted,

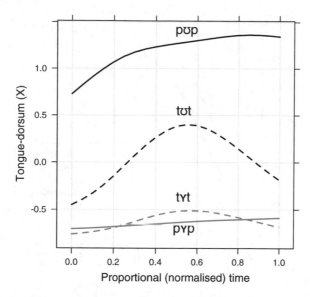

Fig. 6.1 The horizontal position of the tongue dorsum as a function of normalized time extending between the acoustic onset and offset of the /ʊ, ʏ/ (*black, grey*) in /p, t/ (*solid, dashed*) contexts in German non-words. The data are taken from Harrington et al. [56] and were aggregated across seven speakers of German following speaker normalization using z-score [95] normalization. Higher/lower (increasingly positive/negative) values on the y-axis are increasing back/front positions respectively of the tongue dorsum

as for /ʊ/). The results in Fig. 6.2 show the responses from listeners who carried out a forced choice test i.e. identified each stimulus as one of the words. As Fig. 6.2 shows, listeners were more inclined to perceive /u/ in the fronting (*yeast-used*) context. Presumably, this is because they attributed some of the coarticulatory fronting to the consonantal context itself and so biased their responses towards /u/. Another way of putting this is to say that listeners factored out from the acoustic signal the part that was caused by coarticulatory fronting and associated or *parsed* it with the source that gives rise to it, the consonantal context. More generally, listeners of speech have to associate or parse the acoustic consequences of the speaker's interleaved or 'shingled' movements: for the example in Fig. 6.2, they are interleaved because the action of tongue fronting due to the consonantal context is produced or overlaid on the motor actions that are required for the production of the vowel. Models such as articulatory phonology [21] and its forerunner action theory [36, 39] are founded on the premise of a parity between the production and perception modalities: that is, listeners' parsing of the coarticulatory information (/u/-fronting) with the source that is responsible for it (an anterior consonant) is a consequence of their direct perception of the interleaved or shingled movements produced by the speaker.

Ohala's [111, 112] insight is that occasionally the listener does not parse the speech signal in relation to phonological units consistently with their production. A well-known example is coarticulatory nasalization in VN (vowel + nasal

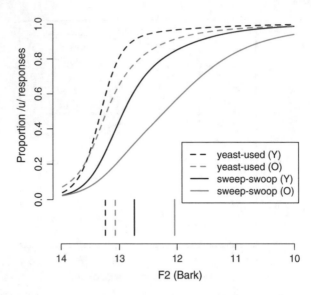

Fig. 6.2 Psychometric curves showing decisions of younger (*black*) and older (*grey*) listeners to continua synthesised between /i, u/ and embedded in fronting (*yeast-used*) and non-fronting (*sweep-swoop*) contexts. The vertical axis shows the proportion of responses identified as /u/ as opposed to /i/. The horizontal axis extends between acoustically most /i/-like (*left*) to acoustically most /u/-like (*right*) vowels. The *vertical lines* mark the cross-over boundaries i.e. where responses were equivocal between /i, u/. Adapted from Harrington et al. [56]

consonant) sequences. If speakers and hearers agree on how the signal is to be parsed into phonological units, then a vowel that is nasalised because it is produced in the context of nasal consonants should be perceived to be oral. Just this has been demonstrated experimentally (e.g. Kawasaki [75]). The interpretation of such findings is as follows: listeners perceive a nasalised vowel in a word like *ban* to be just as oral as in *bad* because they parse (or factor out) the contextual nasalisation in the former and attribute it to the source /n/ (see also Beddor and Krakow [14]).

Such would be the interpretation on the assumption of parity between speech production and perception in the processing of coarticulation. However, a set of perception experiments by Beddor [12] shows that listeners are not always consistent in parsing the coarticulation with the source. If they parse only some of the nasalisation with the following N in VN sequences, then they would hear the vowel to be at least partially nasalised. According to Ohala [112], the historical development of vowel nasalisation in languages such as French (e.g. *main*, 'hand', /mɛ̃/ from Latin *manus*) can arise from just this kind of parsing failure that causes some of the nasalisation to be stuck with the vowel (causing it to be perceived as nasal).

Sound changes like the evolution of French /mɛ̃/ from a VN sequence in Latin is in Ohala's model one of *hypo*correction because the listener has not parsed enough of the coarticulatory variation with the source. A sound change such as the insertion of intrusive /p/ in surnames like *Thompson* derived originally from the *son of Thom* (cf. also *glimpse* from Old English *glimsian*) comes about in terms of this model because the acoustic signal in such nasal clusters often contains a silent interval that listeners incorrectly identify as an oral stop. The silent interval between a nasal consonant /m/ and the fricative /s/ is produced for aerodynamic reasons. An /m/ has an oral movement of lip-closure synchronised with a lowered soft palate in which air exits the nasal cavity. But if the soft palate is raised before the closed lips are released, then the /m/ will change into an oral stop, i.e. a /p/ or /b/, which is marked acoustically by a silent interval (with no air exiting the nasal cavity because the soft palate is raised and none exiting the oral cavity because of the lip-closure). The early raising of the soft palate may come about in order to build up sufficient air pressure in the mouth cavity that is in turn required for the production of the following fricative /s/ with a turbulent airstream. Whatever the reason, a silent interval often occurs between nasals and fricatives: this is why English *mince*/*mints*, /mɪm(t)s, mɪmts/ and German *Gans*/*ganz* ('goose'/'quite') /gan(t)s, gants/ are homophonous for most speakers. From the listener's point of view, the permanent sound change is due in Ohala's model to a parsing failure. There could have been no /p/ in the original production of *Thompson* (being derived from the son of *Thom*) but listeners nevertheless insert one because they cannot parse the silent interval with the /m/ or /s/ as neither of these is typically produced with an acoustically silent interval.

Sound change in Ohala's model can also come about due to *hyper*correction in which the listener parses too much coarticulation with the source. Hypercorrection is presumed to be involved in sound changes such as Grassman's law by which aspiration in ancient Greek came to be deleted if the word contained another aspirated consonant (e.g. /tʰriks/, 'hair' but /trikʰos/ 'hair' gen. sing.). In Ohala's model, this sound change of dissimilation (see e.g. Alderete and Frisch [2], Blevins [15], Müller [104]) comes about because listeners incorrectly parse aspiration in the /tʰ/ of what was presumably an earlier form /tʰrikʰos/ with the following /kʰ/ and so factor it out, in much the same way that listeners typically *correctly* factor out nasalization from the vowel in *ban* and parse it with the /n/ (leading to the perception of an oral vowel in *ban*). Although this theory of perceptual dissimilation is plausible and testable, it has so far been difficult to substantiate it by recreating the conditions under which it could have taken place in the laboratory [1, 58, 104]. Harrington et al. [58] provided some evidence that dissimilation sound changes might be explained by the interaction between coarticulation and speaking style in perception. They showed that long-range coarticulatory lip-rounding can in the perception of a hypoarticulated speaking style (see below) mask the perception of a consonant like /w/ that is inherently lip-rounded thereby possibly recreating the synchronic conditions for its diachronic deletion.

Finally, certain types of metathesis in common with dissimilation can occur when the temporal influence of a sound is extensive. In metathesis, two sounds swap their serial position as in modern English *bird* from Old English *bridde* (see Blevins and Garrett [16] and most recently Egurtzegi [32] for copious other examples). Metathesis often involves liquids (i.e. /l, r/) whose acoustic effects are known to have a long time window [64] i.e. to extend often at least throughout the word, perhaps thereby making the identification of their serial position in relation to the other consonants and vowels of the word difficult for the listener. Just this argument has recently been used in

Ruch and Harrington [128] in modelling the change by which post-aspiration has recently developed from pre-aspiration in the Andalusian variety of Spanish (see also Torreira [147]). In Andalusian Spanish, syllable-final /s/ in words like *pasta* is produced not as an /s/, but is debuccalised and has a quality similar to /h/ in some languages, thus [pa^{h}ta]. Older speakers tend to produce [pa^{h}ta] whereas younger speakers are far more inclined to produce [pat^{h}a] in which the aspiration follows the /t/ (as it does in English *ten*). Ruch and Harrington [128] suggest that this sound change in progress may come about because, while listeners are focussed on whether or not aspiration has occurred (in order to distinguish *pasta* 'pasta' from *pata*, 'paw'), they are unable to determine its serial location in relation to the /t/.

There have been several developments to Ohala's model in recent years including in particular the following.

i. *Small articulatory changes can have large acoustic consequences.* This is the basis of the quantal theory of speech production [140, 141] according to which the relationship between speech production and the acoustic signal (and therefore perception) is non-linear. For example, the incremental retraction of the tongue tip towards the hard palate starting from an /s/ is not accompanied by a similarly gradual acoustic change: instead there is a quantal jump (of spectral centre of gravity lowering of the fricative noise in this case) and an abrupt switch in perception from /s/ to /ʃ/. In a recent ultrasound study of the conditions under which /l/ vocalises—resulting synchronically in productions such as London Cockney /waʊ/ for *wall* and diachronically in /l/-deletion (e.g. in *folk, palm, talk* etc.)—Lin et al. [91] observe that an incremental shift of tongue tip lowering can cause quite a marked acoustic change (of the formant frequencies). Incremental variation in speech production is of course typical (speakers never produce the same utterance exactly identically on two occasions). Perhaps then it is the type of incremental variation that can make a quantal acoustic and perceptual difference which is more likely to evolve into sound change.

ii. *Implementational features* according to Solé [138] facilitate a contrast. For example, a short nasal segment can be produced prior to a voiced stop in Spanish in order to facilitate the production of vocal fold vibration. More specifically, voicing can easily be extinguished in /b, d, g/ because the closure can cause the air pressure in the mouth to approach that in the lungs (below the vocal folds) which, for aerodynamic reasons, would cause the vocal folds to stop vibrating. The production of a preceding brief nasal before the voiced stop reduces the air pressure in the mouth (by channelling the air through the nose) so that the conditions for vocal fold vibration are once more met. Solé [138] shows how this type of implementational feature could have led to the historical development of prenasalised /^{m}b, ^{n}d, ^{n}g/ from voiced stops /b, d, g/ in e.g. Austronesian, Papuan, and South American languages [137]. She also suggests that sound change may often involve just such implementational features, partly because their use varies so much both across and within speakers making it difficult for listeners to parse them with the source (the voiced stop).

iii. *Trading relationships*. In speech production, there are typically multiple acoustic cues for communicating a particular contrast. When two cues are in a trading relationship, then the strength of one cue can vary inversely with the other. For example, /p/ is distinguished from /b/ by a long voice-onset-time (VOT) and/or high-amplitude aspiration noise. In general, the cut-off at which listeners stop hearing /p/ and start to hear /b/ depends on a trading relationship between these cues: the longer the VOT, the lower the amplitude aspiration noise must be (and vice-versa) in order for the cut-off point to remain the same [126]. For Beddor [12], the development of a trading relationship is fundamental to explaining how certain types of coarticulation can evolve into sound change. To return to the VN sequence considered earlier, before the sound change takes hold, listeners factor out nasal coarticulation from the vowel, as described above. However, at some point from synchronic variation to sound change, nasalisation in the vowel and the following nasal consonant are presumed to enter into a trading relationship such that listeners no longer necessarily parse contextual vowel nasalisation with its source (a following N), but instead perceive nasalisation either from information in the vowel or from the following nasal consonant. It is this perceptual change from initially factoring out coarticulation to deploying it in a trading relationship which can begin to provide an explanation for why the source is often deleted as the sound change takes hold [56]. This is because, the more listeners depend on nasal coarticulation in the vowel (to identify nasalisation in a VN sequence), then the less they depend on acoustic information in the following nasal consonant (since the cues are in a trading relationship); thus the extreme case in which nasalisation is perceived entirely in the vowel gives rise increasingly to the perceptual extinction of the source, if the cues are in a perceptual trading relationship. There has been a long-standing puzzle in historical linguistics concerned with how a sound change such as the developing of vowel nasalisation (or indeed umlaut) could result from coarticulation if the source that gives rise to it is subsequently deleted—because if the source wanes then so too should the coarticulatory effect as a result of which the conditions for sound change to take place would no longer be met (see Janda [71] for a further discussion). The idea that the perceptual enhancement of coarticulation may be coupled with this weakening of the source if these cues are in a trading relationship may begin to provide an answer to this puzzle that is grounded in the mechanisms of perception (and how they are related to production).

iv. *Lexical frequency*. Statistical properties of the lexicon and in particular the frequency with which words occur in the language are for some [24, 28, 119, 120] central factors in explaining why sound change occurs. This issue touches upon a long-standing debate of whether, following the Neogrammarian principles [118], sound change spreads through all words of the lexicon at the same rate or whether instead it takes hold first in more frequent words [131, 154]. One of the main reasons why the association between word frequency and sound change is complex is because high frequency words show durational shortening and are often spatially reduced compared with low frequency words [38, 91, 161]. Thus it is not clear whether lexical frequency makes a contribution to sound change independently of a more casual speaking style which also tends to be temporally and spatially reduced. The recent study by Lin et al. [91] points, however, to a rather more direct role of lexical frequency in providing the conditions for sound change to occur: their ultrasound study shows that the tongue tip of /l/ is more likely to be lenited in more frequent words like *milk* than in less frequent words like *elk* but importantly without the /l/ being spatially more reduced due to changes in speaking style. Independently of this finding, an investigation by Zellou and Tamminga [164] showed more extensive nasal coarticulation in more frequent English words in the absence of any durational reduction. The importance of these studies lies in demonstrating therefore that lexical frequency is an independent factor that may predispose sound change to take place. Moreover, given the finding in Lin et al. [91], it seems plausible, as Zellou and Tamminga [164] suggest, that the increased nasal coarticulation in lexically more frequent words is also accompanied by a greater lenition of the nasal stop closure: this would imply that, in frequent like *man*, there is extensive nasalisation in /a/ coupled with tongue tip lenition in the production of /n/. Thus lexically more frequent words may provide the conditions not just for increased coarticulation but also for greater lenition and the subsequent deletion of the source that gives rise to contextual nasalisation (leading to the type of sound change discussed above in which French /mɛ̃/ evolved from Latin *manus*).

v. *Hyper- and hypoarticulation.* According to Lindblom et al. [94], sound change derives from the adaptation of everyday speech communication to the needs of the listener. In this model, speech is produced with extensive spatial and duration reduction (hypoarticulated speech) based on the speaker's prediction that the listener can compensate for the resulting unclear speech by bringing to bear knowledge about the structure of the language. In hypoarticulated speech, the listener's attention is typically not focused on the signal (which often is lacking in clarity) but instead on the content (the semantics of the utterance). If exceptionally the listener processes the phonetic details of hypoarticulated speech, then new forms can be suggested that are added to the listener's lexicon: this is one of the main ways in which reductive sound change (such as 'chocolate' now produced with two syllables /tʃɒklət/) could be added to the lexicon. If coarticulation increases under hypoarticulated speech, then such a model would be able to explain many of the coarticulation-induced sound changes discussed earlier. However, it is not at all clear that coarticulation really does increase in a hypoarticulated speaking style (e.g. Matthies et al. [101], Bradlow [20]). The recent experiments in Harrington et al. [57, 59] and Siddins et al. [134] suggest that hypoarticulation does not magnify coarticulation in production, but it does degrade the listener's ability to factor out the influence of coarticulation: that is, under hypoarticulated speech, there is a more ambiguous relationship between coarticulation and the source that is responsible for coarticulation—just the condition according to which sound change should take place, according to Ohala's [111, 112] model.

vi. *Mismatch between production and perception dynamics*. Harrington et al. [54] and Kleber et al. [82] investigated the relationship between the production and perception of coarticulation for the same speakers during a sound change in progress. The results in Kleber et al. [82] suggest that the association between the perception and production of coarticulation becomes unstable during an ongoing sound change, such that the two modalities are out of alignment (in those typically younger subjects who participate in the sound change) and in which changes to the coarticulatory relationships in perception lead those in production. These findings were predicted from Ohala's [112] model according to which sound change originates initially from the listener's failure to parse coarticulatory dynamics in accordance with the way in which they are produced.

Taken together, points i. and iv. can begin to provide a model for the conditions which predispose sound change to occur that is not functional or teleological (planned). Under a functional view, sound change occurs with the aim of maintaining or even enhancing meaning contrasts. Such a view is central to Martinet's [100] explanation of so-called vowel chain shifts in which the diachronic change in the position of a vowel can have a knock-on effect such that vowels push or pull each other around the vowel space. One of the most recent and striking examples of what seems to be a vowel chain shift is in New Zealand English in which in just the last 50–60 years the front vowels have rearranged themselves such that /a/ → /ɛ/ → /ɪ/ and in which /ɪ/ has centralised: for this reason, a New Zealand English production of *dad* sounds like *dead* and *desk* like *disk*, while *disk* has a vowel quality that is perceived to be not too dissimilar from /ə/ for English speakers of many other varieties [63, 98, 156]. A functional interpretation in Martinet's [100] terms would be that the raising of /a/ to /ɛ/ *causes* the /ɛ/ to raise to /ɪ/ *in order that* the contrasts can be maintained between the three vowels. A functional interpretation is implicit in the model of Lindblom et al. [94] in v. above because sound change arises out of a strategy *in order to* enhance or reduce contrasts based on a prediction of the listener's knowledge.

On the other hand, there may be enough in how the mechanics of speech production, speech perception and the structure of the lexicon are connected without the need to resort to functional explanations. For example, since variation is more likely in lexically frequent words (e.g. Aylett and Turk [3], Wright [161]), then the type of articulatory variation leading to a quantal acoustic change sketched in i. should also be more probable in lexically frequent words. A quantal acoustic change may be one of the factors that contributes to the greater tendency to unlink the coarticulatory variation from the source—to parse nasalisation with the vowel for example in VN sequences. Moreover, the source may be prone to disappear in lexically frequent words because, quite apart from the potential contribution of perceptual trading relationships to sound change (point iii.), the source is reduced (point iv. above) in frequent words and may therefore be less perceptible than in less frequent words. This would complete (or phonologise) the sound change by which in this example nasalisation is associated with the vowel together with deletion of the nasal consonant. Notice how all such principles draw upon naturally occurring phenomena in the production and perception of speech without the need to resort to a functional explanation.

It may also be possible to invoke the same machinery to explain why phonologisation can lead to enhancement [69]. Consider that the vowel /y/ which, in words like *Füße* ('feet'), developed from an /o/ or /u/ in /fotiz/ under the coarticulatory influence of the following /i/, has a much more front quality than would be predicted by synchronic vowel-to-vowel coarticulation. That is, Öhman's [110] study showed that the vowels influence each other but the tongue dorsum is not pushed as far forward as it is in modern German /y/ by the coarticulatory influence of $V_2 = $ /i/ on $V_1 = $ /u/ or /o/. Similarly, while there are phonetic reasons for the vowel to be shorter before a following voiceless than voiced consonant [27, 89], the magnitude of this shortening before voiceless (e.g. *bus*) compared with voiced (*buzz*) consonants in English is far greater than would be predicted by phonetic shortening alone (Ohala and Ohala [114]; see also Solé [136]). Kirby [78, 79] has recently developed a computational model in which phonologisation is an emergent consequence of a combination of precision loss and enhancement. More specifically, the model in Kirby [78] suggests that, as one cue for distinguishing between phonologically different categories wanes, another will be enhanced to ensure that the categories remain distinct. The cue that is enhanced is not necessarily the one that, as in Kingston and Diehl [77], combines with other cues in such a way to enhance an existing phonological contrast (e.g. lip-rounding with tongue backing in the case of /u/), but is instead whichever cue is likely to lead to the greatest probabilistic separation between phonological categories: in Kirby's computer simulation, the cue that comes to be enhanced depends on its acoustic effectiveness for separating between phonological categories, combined with the degree to which it is of use in distinguishing between items in the lexicon. The computer simulation is able to model speech data presented in Kirby [78, 79] showing how fundamental frequency has taken over from duration as the main cue in separating /CrV, CV, ChV/ in the Phnom Penh variety of Khmer.

It is tempting to conclude (as suggested by Kirby's model) that enhancement may be a consequence of phonologisation: speakers enhance the distinctions *because* they are functional i.e. provide an acoustic sharpening of the sounds that are involved in contrasting meaning. Alternatively, consider as discussed earlier that phonologisation often involves the attrition or deletion of the source (of e.g. the nasal consonant) that gives rise to coarticulation (e.g. the nasalisation in the preceding vowel). If as explained in iii. above, the source (nasal consonant) and coarticulatory effect (vowel nasalisation) are in a perceptual trading relationship, then source deletion implies an extreme or at least progressively increasing form of coarticulation, since by definition the two cues are inversely proportional (if they are in a trading relationship). Thus enhancement of coarticulation leading to phonologisation need not be functional, but may be the outcome of what happens when the coarticulatory effect and source enter into a (progressively one-sided) perceptual trading relationship, leading to enhanced coarticulation combined with source attrition.

6.2   Sound Change and Experience

Developments in the last 10–15 years in the episodic or exemplar model of speech perception and production [44, 73, 109, 120] give much greater emphasis to the way in which differences between speakers in their learned phonetic knowledge can contribute to the conditions for sound change to occur. Models of sound change— and indeed more generally speech communication—within an exemplar paradigm are necessarily social: from this point of view, it makes little sense to develop models of sound change that are formulated independently of speaker or listener variation. An exemplar model applied to sound change goes some way towards providing a common framework for models such as Ohala's [111, 112] concerned with the conditions that give rise to sound change and those within the sociolinguistic tradition [85] whose central focus is more on how sound change spreads within a community.

The central idea in an exemplar model of speech is that listeners store in memory episodes i.e. unsegmented auditory gestalts of the words that they hear in a high dimensional perceptual space. Phonological units—that is the abstract units whose permutations function to distinguish between words—emerge from regions of high density in this multidimensional space [121, 122]. Since, for example, all words beginning with /t/ share some basic acoustic properties (of e.g. a rising spectrum in the burst, a second formant frequency that converges around a locus close to 1800 Hz etc.), then the unsegmented auditory traces of such words converge around the (auditory transformation) of these acoustic characteristics: it is such dense regions of intersection that provide the conditions for abstraction i.e. for learning that there is an abstract category such as /t/. Moreover, it is not just phonological information that emerges from stored episodes but also dialect and sociolinguistic information, as well as other speaker characteristics such as gender and age (see Docherty and Foulkes [31] for a further discussion).

The following are some of the most important ways in which exemplar theory has contributed to, or could provide an explanation for, how synchronic variation and sound change are connected.

i. *Coarticulation.* As discussed earlier, Ohala's model suggests that the conditions for the occurrence of sound change are met if listeners misperceive coarticulatory relationships. One of the reasons why they might do so is because the speech production-perception link is non-linear: that is, there are certain kinds of coarticulatory overlap in speech production that are poorly or ambiguously transmitted in speech perception. However, exemplar theory provides at least another reason: if the association between speech and phonological units is idiosyncratic at the level of the individual, then listeners even of the same speaking community may not agree on how to parse coarticulation. For this reason, and as Baker et al. [7] suggest, variation between speakers may make normalising for the effects of coarticulation more difficult. There is, moreover, evidence that listeners parse coarticulation in the same signal differently. For example, Fowler and Brown [37] and Beddor [12, 13] have shown how the extent to which listeners parse nasal coarticulation in the vowel with the following nasal in VN sequences is listener specific. Stevens and Reubold [143] demonstrate the listener-specific way in which pre-aspiration is parsed either with the preceding vowel or with the following geminate in Italian. Yu [162] has extended such findings to show that parsing coarticulation with the source that gives rise to it is affected by the listener's personality and social profile including the extent of autistic-like traits. Harrington et al. [54] and Kleber et al. [82] have both demonstrated perceived coarticulation differences at the group level (see also Kataoka [74]): for the sound change in progress by which /u/ has become fronted under the coarticulatory influence of preceding fronting consonants, older and younger listeners were shown to differ in the extent to which they normalise for coarticulation (Fig. 6.2). Baker et al. [7] have proposed that the conditions for sound change to occur depend on some speakers who produce coarticulation in a particularly exaggerated way—it is these speakers that may be more likely to be imitated (see iv. below) leading to the sound change to be propagated through the community. Baker et al. [7] reason that sound change may be rare (in relation to the ubiquity of synchronic variation) not just because listeners compensate in perception so effectively for context in the way suggested by Ohala, but also because of the scarcity of individuals who exaggerate coarticulation to such an extreme degree. Another reason why sound change may be rare is because only a small number of listeners may respond to and imitate the novel variants that occur in producing exaggerated coarticulation [47].

ii. *Lexical frequency.* More frequent words necessarily give rise to more episodes because by definition we hear them more often. For this reason, the association between phonological categories and speech communication is skewed by statistical properties of the lexicon in an exemplar model. An analysis by Hay and Foulkes [61] of archival recordings of New Zealand English has shown that the progression of the reductive sound change of domain-final /t/-deletion is faster in lexically frequent words. Reubold and Harrington [127] provide some preliminary evidence of an association between sound change and lexical frequency in the same individual. Part of their study was concerned with a longitudinal analysis of the broadcaster Alistair Cooke over a 70-year period. They showed that, having acquired aspects of General American after emigrating to the United States in the 1930s, his accent in later life was becoming again closer to the Received Pronunciation that he produced prior to emigrating. They also provided evidence that this reversion to Received Pronunciation (involving a backing of the vowel in the lexical set BATH) had taken place at a faster rate in lexically frequent than infrequent words. On the other hand, while Labov [88] finds no effect of lexical frequency in vowel chain shifts, Hay et al. [63] show that vowel chain shifts in New Zealand English are led by lexically *infrequent* words because, according to their theory, infrequent words are less likely to have an impact on long-term memory when they occur in regions of vowel overlap.

iii. *Incrementation.* For the Neogrammarians, regular sound change was incremental and not perceptible. In more recent times, the issue of whether regular sound change is incremental or not is more controversial. Thus for some (e.g. Ohala [112], Baker et al. [7]) regular sound change involves an abrupt change between different phonetic variants while for others [15, 56, 83, 102], aspects of regular sound change can be incremental. Irrespective of the controversy, incrementation is predicted by exemplar theory to the extent that sound change comes about through updating a cloud of remembered episodes. It is because the cloud itself is a statistical generalisation over a very large number of episodes that a sound change will initially have only a very small impact, in much the same way that outliers have a small influence on statistical parameters like the mean and variance of a tightly fitting Gaussian probability distribution. But as more and more such outliers accumulate, then the distribution will shift (but incrementally).

iv. The puzzle of the *actuation of sound change* (Weinreich et al. [158]; see also Stevens and Harrington [142] for a recent review) which is concerned with why sound change should take hold at a particular time in one dialect or language but not in another has an explanation in terms of exemplar theory in which the association between episodes and categories, being based on usage and experience, is probabilistic. Independently of sound change, exemplar theory predicts that languages or dialects are likely to differ for similar sets of phonological contrasts: for example, although very many languages contrast voiced /b, d, g/ and voiceless /p, t, k/ stops, there are no known two languages that do so in quite the same way [123]. Just this is to be expected if phonological categories emerge stochastically from speech signals transmitted between speakers and hearers. Analogously, the unstable conditions of the kind reviewed in Sect. 6.1. that could lead to sound change may obtain in one dialect or language but not another, given the presumed probabilistic association between speech signals and phonological categories.

v. *Changes over the lifespan.* This issue (which is discussed in more detail under Sect. 6.4) is concerned with the way in which adults' pronunciation has been shown to change incrementally often over several decades (e.g. Harrington [50], Quené [125], Sankoff and Blondeau [130]) in the direction of changes that have been taking place in the community [53]. The prediction from exemplar theory is that the shift should be dependent on experience. For example, assuming additionally a model in which first and second language acquisition share the same phonetic space, an exemplar model predicts findings such as those in e.g. Sancier and Fowler [129] of a change in a bilingual Portuguese-English adult's productions of American English plosives in the direction of the (shorter) voice onset time of Portuguese, after spending several months in a Portuguese speaking environment in Brazil.

vi. *Category broadening and narrowing.* If, as exemplar theory suggests, categories are derived from absorbing episodes of speech signals that a listener encounters, then categories will evolve into ones that are infinitely broad (due to variation) as a result of which there would be a complete collapse of contrast. The mechanism by which category distinction and stability are maintained could be some form of averaging of episodes in speech production [120]: that is, all episodes are absorbed in perception, but production is based on an aggregate of episodes. This aggregate would create its own episode that, for reasons of statistical sampling, would typically be close to the centre of the distribution. In this way, production, if based on an aggregate, would strengthen the mean of the category; this would, in turn, counteract a category's unlimited broadening. Thus, any phonological category is acted upon by two forces: one that creates a broadening of the category by the uptake of new episodes that are at the probabilistic edges of the distribution; and one that creates a contraction towards the centre through averaging episodes in speech production. It is when one of the edges comes into contact with the distribution from another category that the potential for sound change exists. This is because the edge tokens (for example /e/ vowels with a particularly high F1) will be absorbed into the other category (e.g. /a/). Since tokens are lost at one of the edges, then the force pulling the category in that direction will be weakened, and as a consequence the category will shift slightly in the *opposite* direction (/e/ would shift towards lower F1 values in this example). This change in the balance between opposing forces is simulated in a computational model by Blevins and Wedel [17]. They use the model to explain why the sounds of words whose meaning is not otherwise resolved by context tend not to merge diachronically in just the conditions that would give rise to a loss of contrast i.e. homophony (see their paper for numerous examples of anti-homophony from various languages). Such a model in which a shift in the balance between the strengthening force at the centre and the broadening force at the edge causing categories to repel each other when they are in close proximity could form the basis of explaining vowel push-chains of the kind that were detailed earlier for New Zealand English.

vii. *Sound change through not updating categories*. Exemplar theory suggests that sound change can come about if phonological categories are either not, or only selectively, updated by episodes. In Silverman's [135] analysis of the Mexican language Trique, a sound change has developed by which a /ug, ud/ contrast has evolved into /ug^w, ud/ i.e. with lip-rounding on /g/ but not on /d/. The lip-rounding evidently originates from the lip-rounded vowel /u/, but the issue is why /g/ but not /d/ should become rounded. The crucial insight here is that lip-rounding and /g/ are in a sense acoustically additive so that a lip-rounded /g^w/ is *further away* acoustically and perceptually from /d/ (e.g. Halle et al. [49]), whereas a lip-rounded /d^w/ would cause a lowering of its burst's spectral centre of gravity resulting in an acoustic shift *towards* /g/ (given that /d/ is distinguished from /g/ by high frequency energy in the spectrum; see Harrington [51] for further details). The evolution of this sound change is modelled in terms of rejecting exemplars in regions of ambiguity between categories. Thus, according to this model, a listener is more likely to reject unrounded /ug/ tokens (because these are closer to /ud/ than is /g^w/) and is similarly more likely to reject /ud^w/ tokens (because these are closer to the /g/ distribution than to /d/). In this way, a gradual bias is introduced over successive generations of speakers by which the progressive rejection of /ug/ and /ud^w/ results in the emergence of an acoustically enhanced contrast /g^w, d/. Notice how this is enhancement without teleology: that is, the progressive emergence of the more distinct /g^w, d/ is a natural consequence of rejecting tokens i.e. not updating phonological categories in regions of ambiguity. Similarly, Hay et al. [63] show how a vowel push-chain in New Zealand English (of the kind described in i.) can emerge through not updating categories in regions of ambiguity i.e. ones in which the meaning is compromised. A computer model based on a similar principle is discussed in Garrett and Johnson [42] who also note that not updating categories with ambiguous tokens is consistent with Labov's [85] idea that misunderstood tokens are not used in speech production.

Boersma and Hamann [19] are also concerned with developing a model of sound change in which phonological distinctiveness emerges iteratively through non-purposeful interactions between teachers and learners. They are critical of the mechanisms in (vi., vii.) to maintain stability by having to resort to concepts such as an aggregate across a large number of exemplars in production [120]. Their model adopts by contrast from optimality theory the idea that phonological categories are associated with the speech signal by means of (some 322) constraint-ranked auditory filter bands. The interesting aspect of this model is that stability (i.e. no change) is a compromise between a minimisation of articulatory effort and minimisation of perceptual confusion (which is reminiscent of Lindblom et al. [92] model discussed earlier): sound change is likely to come about when these articulatory and auditory minimisations are not balanced. However, this model has at least as many and perhaps more artificial mechanisms than those proposed within the exemplar paradigm. These include the (so far undemonstrated) idea that children learn an OT-like constraint ranking by bringing into contact their correct knowledge of the underlying phonological structure of words with adult-like acoustic distributions of the phonological categories (a position which Boersma and Hamann admit is unrealistic); and that categories that are auditorily more peripheral are harder to produce (see also Kirby [78] for a similar criticism).

6.3 Sound Change and First Language Acquisition

The idea that there might be direct parallels between language acquisition and sound change has a long history [45, 70, 118, 146] and was taken up in many of the generative and natural [145] phonological frameworks of more recent times (e.g. Lightfoot [90], Kiparsky [80]—see Foulkes and Vihman [34], Beckman [11]; and Diessel [30] for a review). But there are, as shown in Vihman [153] and more recently Foulkes and Vihman [34], serious difficulties with any assumption of such a direct link between the errors produced by children and the forces that give rise to sound change. Firstly, the earlier assumptions about the relationship between acquisition and change are based on treating children as a homogeneous entity when in fact there are differences in the rate of acquisition that can be linked to the development of the lexicon (e.g. Beckman et al. [10], Munson et al. [103]). Secondly, and based on auditory analyses of children's misarticulations in five different languages, Vihman's [153] study shows that many of the typical misarticulations produced by children such as consonant harmony, the simplification of consonant clusters to single consonants, and the shortening of long words are quite uncommon in sound change. Thirdly (and most importantly) it is not enough, as Foulkes and Vihman [34] argue, just to list that there may be parallels between sound change and misarticulation by children (which in any case has not been demonstrated): it is instead necessary to understand the cognitive and social mechanisms by which children's misarticulations or misinterpretations of the speech signal may evolve into diachronic change.

A more fruitful approach towards understanding the connections between child language acquisition and sound change has recently been presented in Beckman et al. [11] who seek to demonstrate that the two types of sound change that are commonly referred to in the sociolinguistics literature as 'from below' and 'from above' the level of awareness interact differently with patterns of speech acquisition in the Seoul variety of Korean and in two varieties of Mandarin. The distinction between sound change 'from above' and 'from below' [85, 87] is not especially well defined, partly because what is above or below awareness or consciousness is not easily integrated with cognitive models of speech processing, and also because the distinction is confounded with social class (changes from a lower-middle class usually being from below; changes from a prestigious class, from above—see Wardhaugh and Fuller [155] for a further discussion). Nevertheless, there is a rough correspondence between sound change from below and the regular (as opposed to analogic) changes that have their origins in the processes of coarticulation and lenition, although sound change from below typically also encompasses a sociolinguistic dimension of less prestigious, more working-class speech that is often found in a local vernacular. Sound changes from below like those due to coarticulation are incremental in that they progress gradually across generations [87]. By contrast sound change from above—which is often brought about by contact between speakers of different dialects or languages—more typically involves an abrupt change from one phoneme or category to another and may be driven by a more prestigious or powerful group of individuals. A possible example of a sound change from above might be the increasing use in Standard Southern British English of high-rising-terminals in which the intonation rises in declarative sentences: there is no sense in which this change is incremental (it was not the case that falling intonation gradually evolved via mid-level intonation into rising intonation over a number of years), it is arguably brought about through dialect contact (with North American and/or Australian varieties) and it is a change that listeners have commented upon (and so cannot be, in Labov's terminology, below the level of consciousness). Notice as well that such a sound change is—like many sound changes from above—idiosyncratic in that, unlike the coarticulation-based sound changes discussed earlier, it is not usually found across multiple languages.

Labov [84, 87] has also argued that sound change from below is often led by women: the association with language acquisition is their typically greater involvement than men as caregivers. For example, Labov [84] proposes that many sound changes from below originate in the lower-middle classes; that for these classes there is the greatest differentiation between male and female speakers (especially since female speakers are most likely to style shift and copy pronunciations of the upper classes in a more formal speaking style); and that consequently, since women are the primary caregivers, infants' speech will be influenced more by women than by men (as a result of which sound change from below is often led by women). This view about which gender leads sound change is, however, complicated by more recent findings showing that speaking style in child-directed speech is differently affected by whether caregivers are talking to infant boys or girls (e.g. Foulkes et al. [35]).

Beckman et al.'s [11] study has made advances to our understanding of how the progress of sound change and language acquisition might be connected. They do so through analyses of adult data over a 60 year period combined with perceptual studies of a sound change by which in Seoul Korean the distinction between lax versus aspirated stops is increasingly cued not as in earlier times by voice onset time, but by fundamental frequency differences. They argue that the change is 'from below': this is because a similar type of sound change is found in many other languages; and, as is typical for a sound change from below, it has progressed at a faster rate in women. They then show that infants are not as advanced in this sound change: that is, infants make more use of VOT than younger women. This comes about, Beckman et al. [11] reason, because their caregivers are from an older generation. Following Hockett [66], it is only when children at a slightly older age mix with peers and other social groups that they will not just imitate but incrementally advance the change further still (see also Kerswill [76]). The other finding in Beckman et al. [11] is that the association with language acquisition is very different for the sound change from above by which a three-way fricative contrast has developed in the Sōngyuán variety under the influence of standard Mandarin. Importantly, children of the Sōngyuán variety do not show any incremental progression but instead copy categorically the three-way fricative contrast that does not exist for adult speakers of Sōngyuán.

A different approach to the relationship between sound change and acquisition that relates rather more directly to Ohala's model (see also Greenlee and Ohala [46]) is explored in Kleber and Peters [81]: their concern is to test whether as less experienced users of the language, children are more likely to have difficulty normalising in perception for context effects such as coarticulation. If this is so, then there may be a greater potential for children to fail to attribute a coarticulatory effect to the source that gives rise to it, thus providing the conditions for sound change to occur. A study by Nittrouer and Studdert-Kennedy [108] provided some evidence that adults normalise to a greater extent for the coarticulatory effects of vowel context on preceding fricatives than children; and that there was also a greater degree of adult-like normalisation in 7-year old compared with 3-year old children. Consistently with this study, Kleber and Peters [81] have found that children are much more variable in perception than adults in normalising for the effects of context. A subsequent perception study by Harrington et al. [58] compared young first language German children and adults on the extent to which they normalised for the coarticulatory influences of /p_p/ and /t_t/ contexts on German high front /ʊ/ (back) and /Y/ (front) vowels. Their results showed that the distance between the decision boundaries on an /ʊ-Y/ continuum were closer together for the children than for adults which suggests that children might have been less sensitive to the perceptual influence of the consonantal context on the vowel. However, the interpretation of adult-child differences in Harrington et al. [58] is not that children normalise less for coarticulation, but instead that they are less certain than adults about phonological categorisation, in a way that is analogous to the greater uncertainty when adult listeners categorise hypoarticulated speech signals [57, 59].

6.4 Sound Change and Imitation

The last few years have seen a focus on spontaneous imitation in speech (Delvaux and Soquet [29], Nielsen [106], Pardo et al. [116, 117], Yu et al. [163] to mention but a few) and some attempts to consider the implications of these findings for models of sound change [6, 42, 52]. The methodology for demonstrating imitation typically involves comparing subjects' speech production before and after they have performed a task such as listening to or shadowing another speaker; imitation has also been measured and demonstrated in terms of the degree of convergence between speakers in a conversation [116]. In general, two main findings have emerged from this research. The first is that imitation can take place without any social motivation to do so. This is so, for example, in Nielsen's [106] study in which imitation is measured from the production of isolated words. This type of spontaneous imitation may be part of a more general tendency for individuals to coordinate their actions in space and time [40, 132] that can give rise to alignment of many different kinds—for example of body movements [133] and of syntactic structures [43]. Secondly, imitation is by no means automatic and inevitable for all speakers [163] and can vary depending on the speaking situations [116]. In addition, imitation can be constrained by the social context including how unusual [6] or attractive [5] the interlocutor's voice is perceived to be, as well as the attitudes of the speaker towards the interlocutor's national identity [4].

In the analyses of the Christmas broadcasts produced annually by Queen Elizabeth II, Harrington et al. [53] showed that the Queen's 1950s vowels had shifted over 30 years towards those of a mainstream Standard Southern British (SSB) more typically produced by the middle classes. The progression which was found for several vowels, was gradual and in the direction of sound change that had been taking place to the standard accent of England over a 50 year period; it was also away from an aristocratic form of the Queen's 1950s variety (referred to by Wells [159] as U- or upper-crust RP) towards a more middle-class variety of SSB. Moreover, the shift was partial such that the Queen's 1980s vowels were generally at intermediary locations between those from the 1950s and those that were more typical of the five analysed mainstream SSB speakers in Harrington et al. [53] who had been recorded in the 1980s.

It seems quite probable that these shifts to the Queen's vowels have been brought about because of some form of imitation or convergence in dialogue. As discussed in Harrington [56], the 1960s and 1970s saw a rise of the lower-middle and middle classes into positions of power, with the result that the Queen is likely to have come into increasing contact in the intervening decades with speakers of a more middle class variety. Indeed, the prime ministers to which the Queen gives a weekly audience were in the 1960s and 1970s (James Callaghan, Edward Heath, Margaret Thatcher, Harold Wilson) generally of more humble origins than their predecessors in the

1950s and (with the possible exception of Heath) with accents that were removed from U-RP of the 1950s.

The question that must be considered is why the Queen should shift her accent towards those speakers with a less aristocratic, more mainstream form of Standard Southern British which would include not just her prime ministers but quite possibly also staff who arguably might have been increasingly likely to produce a less aristocratic form of RP after the 1950s. Perhaps it was because the Queen engaged in style-shifting [62] towards an accent that typified a more egalitarian society that was shaping England in the late 1960s and 1970s. But the reason why this interpretation is unlikely is because the sound change in these broadcasts is gradual. Moreover, the changes to the Queen's vowels are slow (some 10 Hz per annum in the fronting of /u/) which would have repeatedly required quite accurate style-shifting of a few Hz per annum over the thirty year period [52]. A more plausible explanation is that personal accent is incrementally influenced over a long period of time through daily or regular spoken interaction with speakers from another dialect group or indeed from another language [129]. Incrementation is, as explained earlier, also predicted by exemplar theory: the accumulation over time of episodes of speech signals from speakers of other dialect groups should shift incrementally the statistical distributions of a person's phonological categories: that is, interacting with and listening to speakers over 2–3 decades who have a phonetically more fronted /u/ has fronted incrementally the Queen's 1950s /u/ towards a form intermediate between U-RP of the 1950s and SSB of the 1980s.

The positions of the Queen's 1980s vowels between those of her 1950s vowels and those of the five mainstream SSB speakers recorded in the 1980s is predicted by a dialect mixture model in which there has been a blending or averaging of the aristocratic, conservative accent of the 1950s with the more middle-class, mainstream SSB accent of the 1980s.

The above interpretation of incrementation through interaction is also entirely consistent with the model of dialect mixture that is developed by Trudgill [151] in his analyses of New Zealand English. Trudgill argues that the creation of this accent has been a deterministic function of a levelling of the different dialects of the speakers who first came to New Zealand in the 19th century: that is, New Zealand English is not the result of the need for New Zealanders to establish group membership and identities that are unique to their country. For Trudgill [151], the evolution of New Zealand English as a consequence of dialect mixture is due to an 'innate tendency to behavioral coordination' (p. 252) among the interlocutors, a process that is the consequence of interaction. Building on an earlier idea from Bloomfield [18], he notes that this model of dialect mixture is in agreement with Labov's [86] view that in many cases the diffusion of linguistic change can be explained through communication density i.e. as a consequence of the speakers interacting with each other. Importantly, Labov [86] notes that there need not be any motivating social force behind such changes which can be mechanical and inevitable and in which social evaluation plays only a very minor role (Labov [86], p. 19–20). Trudgill's [151] idea of new dialect formation via non-social determinism is a further development of his earlier gravity model in which the spread of a linguistic innovation depends on a combination of geographical distance and population density. Trudgill's [148] gravity model was used to explain how an innovation such as /h/-dropping spreads in England from city to city while skipping over less populated rural areas. As described in further detail by Wolfram and Schilling-Estes [160], the gravity model was subsequently modified to take account of other findings [149] suggesting the likelihood of change is greater when the two dialects are phonetically similar to each other.

This type of dialect averaging has recently been demonstrated for a completely different set of data by Bukmaier et al. [22] in a study concerned with dialect levelling due to contact with a standard variety. The focus of their study is the Augsburg variety of German which, in contrast to standard German, neutralises the /s-ʃ/ contrast in certain pre-consonantal contexts: thus standard German /mɪst, mɪʃt / (misst/mischt; 'measures'/'mixes') are both /mɪʃt / in Augsburg German. Now it might be supposed that any change in the direction of the standard is categorical by which younger speakers typically produce more /s/ in words like misst than their older counterparts. But this is not what the instrumental analysis showed: instead Bukmaier et al. [22] found that younger Augsburg subjects were intermediate between older Augsburg and standard German subjects in the fricative of words like misst. This was so in both production and perception. Thus their produced forms had all the characteristics of a sibilant fricative that was intermediate between /s/ and /ʃ/; and in perception, they were also intermediate between the poor ability of older Augsburg and sharp discrimination of standard listeners in distinguishing between word pairs like /fɛɐ̯'mɪstə, fɛɐ̯'mɪʃtə/ (vermisste/vermischte; 'missed'/'mingled'). There was a similar finding in an apparent time study comparing speakers of East Franconian German with standard German speakers [56, 105]. In East Franconian, as in many other varieties of German, the post-vocalic voicing contrast is neutralised towards the voiced variety: thus leiten/leiden, /laɪtn, laɪdn/ ('to lead'/ 'to suffer') are both /laɪdn/ in the East Franconian variety. However, younger East Franconians were found to produce and perceive a contrast that was intermediate between those of older East Franconians and standard German subjects. Such intermediate positions for both Augsburg and East Franconian German during a sound change in progress due to dialect contact are exactly what is predicted by Trudgill's [151] dialect mixture model.

Trudgill [150] also shows that the first two generations of young children were likely to have been the prime instigators of the dialect mixture leading to the development of New Zealand English; he also suggests [152] that children may be especially prone to the type of interactional alignment described above. Children's accents are of course very malleable: the accent of a child rapidly shifts towards that of its peers and away from that of its parents, which is especially noticeable if they and their families are recent migrants to a different dialect region, as Chambers [26] and others have shown. Recently, Nielsen [107] showed that there was more imitation of (a lengthened) voice onset time in stops by children—both pre-schoolers and those aged 8–9 years than by adults. Consider in addition Babel's [6] recent finding of a predisposition to imitate novel voices. Perhaps then one of the reasons why children may be predisposed to imitate is because of the large number of novel voices that they encounter as the child's social network typically expands beyond that of the family and caretakers. The imitation may be especially likely because, having previously encountered only a few voices and having relatively limited speaking experience, the statistical distributions of their emerging phonological categories would still be quite unstable and therefore strongly affected by novel voices and pronunciations (see Stanford and Kenny [139] for similar views).

On the other hand, computational simulations using agent-based modelling suggest that imitation and non-social determinism—which Trudgill [151] argues are central factors in new dialect formation—may not be sufficient to explain how sound [8] or language [124] change propagate through the community. Agent-based models apply principles from statistical physics to social dynamics in order to understand how local interaction between individuals—represented by agents who are interconnected in a network—can bring about global (community) changes [25]. The agent-based social network model in Fagyal et al. [33] was designed to test various theories about how the inter-connectedness of individuals is related to linguistic change. Their general conclusion is that language change is propagated principally by so-called leaders that is those with many connections to other individuals and that linguistic change will only spread through the community if there are both leaders and 'loners' i.e. those with far fewer social connections. Their model also provides some support for the idea that prestige drives language change. In Pierrehumbert et al.'s [124] agent-based computational model, an individual's choice between alternative linguistic categories is based on three main factors: the community norm modelled as the aggregated input from the other speakers with which the individual is linked in the network; stored knowledge consisting of a bias factor that governs whether a given individual prefers conservative or novel forms; and finally a variable (designated as 'temperature') that controls the degree to which individuals randomly or categorically choose between alternatives (i.e. their model incorporates the idea that individuals select between alternatives with varying degrees of probability). Compatibly with Fagyal et al. [33], their simulations show that a mixture of individuals who are innovators (prone to adopt a form that is not consistent with community norms) and who resist change are prerequisites for language change. But in contrast to Fagyal et al. [33], highly connected individuals tend not to be innovative in Pierrehumbert et al.'s [124] simulations because their output is so strongly influenced by the large input that they receive from other more conservative individuals with which they are connected. Change is instead more likely to radiate from individuals in their model who are innovative, connected to other individuals who are similarly biased towards innovation, but who are only average or below average as far as the total number of connections to other individuals is concerned. Pierrehumbert et al. [124] use this model to suggest that sound change originates from closely-knit communities whose speakers share innovations and also to argue against the idea that prestige is a driving force for sound change. One of the major innovations in this model is that a probabilistic (rather than a binary) choice between categories or variants is one of the components in their model for language change to spread through the community.

6.5   Concluding Comments

Sound change is fascinating and scientifically tractable. There are speech signals
on the one hand and there are categorical changes (*bird* < *bridde*) on the other and
there is some form of cognitive, physiological, and perceptual machinery linking the
two. The task is to assemble many different forms of empirical evidence—ranging
across etymological reconstructions of changing sound patterns between ancient and
modern languages, the social aspects of speech, physiological analyses of movement
and their perceptual consequences—to find out what form the machinery takes in
order to explain how signals and sound change are connected. This task is in turn
illuminating for the most challenging task in the speech sciences: how the categories
of speech—the consonants and vowels out of which words are constructed—and
signals in human speech communication and the machine derivative thereof are inter-
related. This is fundamental knowledge for both human and machine processing.

The review of the current state of the literature has shown that some themes
for explaining the operation of sound change recur in several studies often across
different disciplines. Some of the most important of these are as follows.

   i. *Coarticulation* or more generally the dichotomy between the serial order of
      phonological categories and the interleaved, shingled movements in the speech
      signal remains fundamental to many different types of sound change (assimi-
      lation, dissimilation, metathesis, phonologisation) that recur through different
      languages.
  ii. *Perception and its relationship to speech production.* Many different studies have
      shown how non-linearity between the two modalities can provide the conditions
      for sound change to take place: this is so when coarticulatory dynamics are mis-
      perceived which may also provide the basis for explaining the often asymmetric
      direction of sound change. In addition, coarticulation leading to sound change
      may become perceptually salient when articulatory incrementation pushes the
      acoustic signal across a quantal boundary. Perceptual trading relationships could
      be important in understanding phonologisation and also why phonologisation
      seems to lead to enhancement.
 iii. *Hypoarticulation* which is the condition under which the speech signal is
      degraded often in relation to its predictability from context is important for
      understanding sound change, not just from the perspective that many types of
      sound change are reductive (involving e.g. consonant lenitions—e.g. Bybee [23])
      but also because hypoarticulation combined with a possible degradation of the
      coarticulatory source may obscure coarticulatory parsing in perception [58, 59].
      Hypoarticulation accounts also emphasise how stored (top-down) knowledge—
      such as lexical frequency may be implicated in sound change.
  iv. *Experience* (exemplar) based models of speech have provided numerous addi-
      tional ways and also an appropriate metalanguage for understanding the oper-
      ation of sound change. Experience-based models can explain how speaker
      variation—both at the individual and at the group level—can create the condi-
      tions for sound change to occur. The perception-production feedback loop [157]

inherent in exemplar theory coupled with findings from imitation can explain the occurrence of sound change over the lifespan and the emergence across generations of a mixture when two dialects come into contact. This perception-production feedback loop explains the effects of lexical frequency; it also provides the mechanism for (and predicts) that regular sound change is incremental.

Finally, and as discussed in Blevins and Wedel [17], in much the same way that Darwin's theory of natural selection has led to an understanding of how purposeless interactions at the level of the individual can give rise to what appears to be a purposeful population change, so too can cumulative variation at a number of nested and interlocking levels between individual speakers and a community, and between speech signals and the lexicon, push the sounds of the language between stable and unstable states. This metaphor also provides a way of understanding sound change without having to invoke the untestable (or at least so far undemonstrated) idea that sound change is started by any individual (or pair of individuals as in Ohala's [112] model) and then spreads through the community. It also implies that the sharp division that is made by many between the so-called origin of sound change and its spread (e.g. Ohala [112], Baker et al. [7], Janda and Joseph [72]) is a fallacy. The sounds of language are in a constant state of flux i.e. there is no point at which sound change is not taking place (and so no fixed point at which sound change starts). There are instead multiple conditions that can create instabilities that lead to categorical change. Understanding these through the wide range of theories and empirical techniques that have been reviewed in this paper remains an exciting challenge in the future.

Acknowledgments This research was supported by European Research Council grant number 295573 'Sound change and the acquisition of speech' (2012–2017). We are very grateful to Daniela Müller, two anonymous reviewers, and the editors for very helpful comments.

References

1. Abrego-Collier C (2013) Liquid dissimilation as listener hypocorrection. In: Proceedings of the 37th annual meeting of the Berkeley linguistics society, 3–17
2. Alderete J, Frisch S (2006) Dissimilation in grammar and the lexicon. In: de Lacy P (ed) The cambridge handbook of phonology. Cambridge University Press, Cambridge
3. Aylett M, Turk A (2004) The smooth signal redundancy hypothesis: a functional explanation for relationships between redundancy, prosodic prominence, and duration in spontaneous speech. Lang Speech 47:31–56
4. Babel M (2010) Dialect divergence and convergence in New Zealand English. Lang Soc 39:437–456
5. Babel M (2012) Evidence for phonetic and social selectivity in spontaneous phonetic imitation. J Phon 40:177–189
6. Babel M, McGuire G, Walters S, Nicholls A (2014) Novelty and social preference in phonetic accommodation. Lab Phonol 5:123–150
7. Baker A, Archangeli D, Mielke J (2011) Variability in American English s-retraction suggests a solution to the actuation problem. Lang Var Chang 23:347–374

8. Baxter G, Blythe R, Croft W, McKane A (2009) Modeling language change: an evaluation of Trudgill's theory of the emergence of New Zealand English. Lang Var Chang 21:257–293
9. Beckman M, Edwards J (1994) Articulatory evidence for differentiating stress categories. In: Keating P (ed) Papers in laboratory phonology III. Cambridge University Press, Cambridge, pp 7–33
10. Beckman M, Munson B, Edwards JR (2007) Vocabulary growth and the developmental expansion of types of phonological knowledge. In: Hualde J, Cole J (eds) Papers in laboratory phonology IX. Cambridge University Press, Cambridge, pp 241–264
11. Beckman M, Li F, Kong E, Edwards J (2014) Aligning the timelines of phonological acquisition and change. Lab Phonol 5:151–194
12. Beddor P (2009) A coarticulatory path to sound change. Language 85:785–821
13. Beddor P (2012) Perception grammars and sound change. In: Solé M-J, Recasens D (eds) The initiation of sound change. Perception, production, and social factors. John Benjamin, Amsterdam
14. Beddor PS, Krakow RA (1999) Perception of coarticulatory nasalization by speakers of English and Thai: evidence for partial compensation. J Acoust Soc Am 106:2868–2887
15. Blevins J (2004) Evolutionary phonology: the emergence of sound patterns. Cambridge University Press, Cambridge
16. Blevins J, Garrett A (2004) The evolution of metathesis. In: Hayes B, Kirchner R, Steriade D (eds) Phonetically-based phonology. Cambridge University Press, Cambridge, pp 117–156
17. Blevins J, Wedel A (2009) Inhibited sound change: an evolutionary approach to lexical competition. Diachronica 26:143–183
18. Bloomfield L (1933) Language. Holt, New York
19. Boersma P, Hamann S (2008) The evolution of auditory dispersion in bidirectional constraint grammars. Phonology 25:217–270
20. Bradlow AR (2002) Confluent talker-and listener-oriented forces in clear speech production. In: Gussenhoven C, Warner N (eds) Papers in laboratory phonology VII. Mouton de Gruyter, New York, pp 241–273
21. Browman C, Goldstein L (1992) Articulatory phonology: an overview. Phonetica 49:155–180
22. Bukmaier V, Harrington J, Kleber F (2014) An analysis of post-vocalic /s-ʃ/ neutralization in Augsburg German: evidence for a gradient sound change. Front. Psychol 5:1–12
23. Bybee J (2001) Phonology and language use. Cambridge University Press, Cambridge
24. Bybee J (2002) Word frequency and context of use in the lexical diffusion of phonetically conditioned sound change. Lang Va Chang 14:261–290
25. Castellano C, Fortunato S, Loreto V (2009) Statistical physics of social dynamics. Rev Mod Phys 81:591–646
26. Chambers J (1992) Dialect acquisition. Language 68:673–705
27. Chen M (1970) Vowel length variation as a function of the voicing of the consonant environment. Phonetica 22:129–159
28. Chen M, Wang W (1975) Sound change: actuation and implementation. Language 51:255–281
29. Delvaux V, Soquet A (2007) The influence of ambient speech on adult speech productions through unintentional imitation. Phonetica 64:145–173
30. Diessel H (2012) Language change and language acquisition. In: Bergs A, Brinton L (eds) Historical linguistics of English: an international handbook, vol 2. Mouton de Gruyter, Berlin, pp 1599–1613
31. Docherty G, Foulkes P (2014) An evaluation of usage-based approaches to the modelling of sociophonetic variability. Lingua 142:42–56
32. Egurtzegi A (2014) Towards a phonetically grounded diachronic phonology of Basque. PhD dissertation, University of the Basque Country
33. Fagyal Z, Escobar A, Swarup S, Gasser L, Lakkaraju K (2010) Centers and periph- eries: network roles in language change. Lingua 120:2061–2079
34. Foulkes P, Vihman MM (in press) Language acquisition and phonological change. In: Honeybone P, Salmons JC (eds) The handbook of historical phonology. OUP, Oxford

35. Foulkes P, Docherty G, Watt D (2005) Phonological variation in child directed speech. Language 81:177–206
36. Fowler C (1984) Segmentation of coarticulated speech in perception. Percept Psychophys 36:359–368
37. Fowler C, Brown J (2000) Perceptual parsing of acoustic consequences of velum lowering from information for vowels. Percept Psychophys 62:21–32
38. Fowler C, Housum J (1987) Talkers' signaling of 'new' and 'old' words in speech and listeners' perception and use of the distinction. J Mem Lang 26:489–504
39. Fowler C, Smith M (1986) Speech perception as "vector analysis": an approach to the problems of segmentation and invariance. In: Perkell J, Klatt D (eds) Invariance and variability in speech processes. Erlbaum, Hillsdale, pp 123–139
40. Fowler C, Richardson M, Marsh K, Shockley K (2008) Language use, coordination, and the emergence of cooperative action. In: Fuchs A, Jirsa V (eds) Understanding complex systems. Springer, Berlin, pp 261–279
41. Fujisaki H, Kunisaki O (1976) Analysis, recognition and perception of voiceless fricative consonants in Japanese. Annu Bull Res Inst Logop Phoniatr 10:145–156
42. Garrett A, Johnson K (2013) Phonetic bias in sound change. In: Yu A (ed) Origins of sound change. Oxford University Press, Oxford, pp 51–97
43. Garrod S, Pickering M (2009) Joint action, interactive alignment, and dialog. Top Cogn Sci 1:292–304
44. Goldinger S (1997) Words and voices: perception and production in an episodic lexicon. In: Johnson K, Mullennix J (eds) Talker variability in speech processing. Academic Press, San Diego, pp 33–66
45. Grammont M (1902) Observations sur le langage des enfants. In Me langes Linguistiques Offerts al M. Antoine Meillet. Klincksieck, Paris, pp 115–131
46. Greenlee M, Ohala J (1980) Phonetically motivated parallels between child phonology and historical sound change. Lang Sci 2:283–301
47. Grosvald M, Corina D (2012) The production and perception of sub-phonemic vowel contrasts and the role of the listener in sound change. In: Solé M-J, Recasens D (eds) The initiation of sound change. Perception, production, and social factors. John Benjamins, Amsterdam, pp 77–100
48. Guion S (1998) The role of perception in the sound change of velar palatalization. Phonetica 55:18–52
49. Halle M, Hughes G, Radley J-P (1957) Acoustic properties of stop consonants. J Acoust Soc Am 29:107–116
50. Harrington J (2006) An acoustic analysis of 'happy-tensing' in the Queen's Christmas broadcasts. J Phon 34:439–457
51. Harrington J (2010) Acoustic phonetics. In: Hardcastle W, Laver J, Gibbon F (eds) A handbook of phonetics. Wiley-Blackwell, Oxford, pp 81–129
52. Harrington J (2012) The relationship between synchronic variation and diachronic change. In: Cohn A, Fougeron C, Huffman M (eds) Handbook of laboratory phonology. Oxford University Press, Oxford, pp 321–332
53. Harrington J, Palethorpe S, Watson C (2000) Does the Queen speak the Queen's English? Nature 408:927–928
54. Harrington J, Kleber F, Reubold U (2008) Compensation for coarticulation, /u/-fronting, and sound change in Standard Southern British: an acoustic and perceptual study. J Acoust Soc Am 123:2825–2835
55. Harrington J, Hoole P, Kleber F, Reubold U (2011) The physiological, acoustic, and perceptual basis of high back vowel fronting: evidence from German tense and lax vowels. J Phon 39:121–131
56. Harrington J, Kleber F, Reubold U (2012) The production and perception of coarticulation in two types of sound change in progress. In: Fuchs S, Weirich M, Perrier P, Pape D (eds) Speech production and speech perception: planning and dynamics. Peter Lang, Bern, pp 33–55

57. Harrington J, Kleber F, Reubold U (2013) The effect of prosodic weakening on the production and perception of trans-consonantal vowel coarticulation in German. J Acoust Soc Am 134:551–561

58. Harrington J, Kleber F, Stevens M (2015) The relationship between the (mis)-parsing of coarticulation in perception and sound change: evidence from dissimilation and language acquisition. In: Esposito A, Faundez-Zany M (eds) Recent advances in nonlinear speech processing. Springer, Berlin

59. Harrington J, Kleber F, Reubold U, Siddins J (2015) The implications for sound change of prosodic weakening: evidence from polysyllabic shortening. Lab Phonol 6(1):87–117

60. Hawkins S (2003) Roles and representations of systematic fine phonetic detail in speech understanding. J Phon 31:373–405

61. Hay J, Foulkes P (in press) The evolution of medial /t/ over real and remembered time. Language

62. Hay J, Jannedy S, Mendoza-Denton N (1999) Oprah and /ay/: lexical frequency, referee design and style. In: Proceedings of the 14th international congress of phonetic sciences, San Francisco

63. Hay J, Pierrehumbert J, Walker A, LaShell P (2015) Tracking word frequency effects through 130 years of sound change. Cognition 139:83–91

64. Heid S, Hawkins S (2000) An acoustical study of long domain /r/ and /l/ coarticulation. In: Proceedings of the 5th seminar on speech production: models and data. Kloster Seeon, Bavaria, Germany. Munich, pp 77–80

65. Hintzman DL (1986) 'Schema abstraction' in a multiple-trace memory model. Psychol Rev 93:328–338

66. Hockett C (1950) Age-grading and linguistic continuity. Language 26:449–457

67. Hoole P, Pouplier M, Benus S, Bombien L (2013) Articulatory coordination in obstruent-sonorant clusters and syllabic consonants: data and modelling. In: Spreafico L, Vietti A (eds) Rhotics: new data and perspectives. Bolzano University Press, pp 81–97

68. Hualde J (2011) Sound change. In: van Oostendorp M, Ewen C, Hume E, Rice K (eds) The blackwell companion to phonology, vol IV. Wiley-Blackwell, Oxford, pp 2214–2235

69. Hyman LM (2013) Enlarging the scope of phonologization. In: Yu A (ed) Origins of sound change: approaches to phonologization. Oxford University Press, Oxford, pp 3–28

70. Jakobson R (1941) Kindersprache, Aphasie, und allgemeine Lautgesetze. Almqvist & Wiksell, Uppsala

71. Janda R (2003) "Phonologization" as the start of dephoneticization—or, one sound change and its aftermath: of extension, generalization, lexicalization, and morphologization. In: Joseph B, Janda R (eds) The handbook of historical linguistics. Blackwell, Oxford, pp 401–422

72. Janda R, Joseph B (2003) Reconsidering the canons of sound-change: towards a "big bang" theory. In: Blake B, Burridge K (eds) Selected papers from the 15th international conference on historical linguistics. John Benjamins, Amsterdam, pp 205–219

73. Johnson K (1997) Speech perception without speaker normalization: An exemplar model. In: Johnson K, Mullennix J (eds) Talker variability in speech processing. Academic Press, San Diego, pp 145–165

74. Kataoka R (2011) Phonetic and cognitive bases of sound change. Ph.D diss. University of California, Berkeley

75. Kawasaki H (1986) Phonetic explanation for phonological universals: the case of distinctive vowel nasalization. In: Ohala J, Jaeger J (eds) Experimental phonology. Academic Press, Orlando, pp 81–103

76. Kerswill P (1996) Children, adolescents, and language change. Language Variation and Change 8:177–202

77. Kingston J, Diehl RL (1994) Phonetic knowledge. Language 70:419–454

78. Kirby J (2013) The role of probabilistic enhancement in phonologization. In: Yu A (ed) Origins of sound change: approaches to phonologization. Oxford University Press, Oxford, pp 228–246

79. Kirby J (2014) Incipient tonogenesis in Phnom Penh Khmer: computational studies. Laboratory phonology 5:195–230
80. Kiparsky P (2008) Universals constrain change; change results in typological generalizations. In: Good J (ed) Linguistic universals and language change. Oxford University Press, Oxford, pp 23–53
81. Kleber F, Peters S (2014) Children's imitation of coarticulatory patterns in different prosodic contexts. In: Proceedings of 14th conference on laboratory phonology, Tokyo, Japan
82. Kleber F, Harrington J, Reubold U (2012) The relationship between the perception and production of coarticulation during a sound change in progress. Lang Speech 55:383–405
83. Labov W (1981) Resolving the Neogrammarian controversy. Language 57:267–308
84. Labov W (1990) The intersection of sex and social class in the course of linguistic change. Lang Var Chang 2:205–254
85. Labov W (1994) Principles of linguistic change. Volume 1 internal factors. Blackwell, Malden
86. Labov W (2001) Principles of linguistic change. Volume 2: social factors. Blackwell, Oxford
87. Labov W (2007) Transmission and diffusion. Language 83:344–387
88. Labov W (2010) Principles of linguistic change, Volume 3: cognitive and cultural factors. Wiley-Blackwell, Oxford
89. Lehiste I (1970) Suprasegmentals. MIT Press, Cambridge
90. Lightfoot D (1999) The development of language: acquisition, change, and evolution. Blackwell, Malden
91. Lin S, Beddor P, Coetzee A (2014) Gestural reduction, lexical frequency, and sound change: a study of post-vocalic /l/. Lab Phonol 5:9–36
92. Lindblom B (1990) Explaining phonetic variation: a sketch of the H & H theory. In: Hardcastle W, Marchal A (eds) Speech production and speech modeling. Kluwer, Dordrecht, pp 403–439
93. Lindblom B, Studdert-Kennedy M (1967) On the role of formant transitions in vowel recognition. J Acoust Soc Am 42:830–843
94. Lindblom B, Guion S, Hura S, Moon S-J, Willerman R (1995) Is sound change adaptive? Rivista di Linguistica 7:5–36
95. Lobanov B (1971) Classification of Russian vowels spoken by different speakers. J Acoust Soc Am 49:606–608
96. Local J (2003) Variable domains and variable relevance: interpreting phonetic exponents. J Phon 31:321–329
97. Luick K (1964) Historische Grammatik der englischen Sprache. Blackwell, Oxford
98. Maclagan M, Hay J (2007) Getting fed up with our feet: contrast maintenance and the New Zealand English "short" front vowel shift. Lang Var Chang 9:1–25
99. Mann V, Repp B (1980) Influence of vocalic context on the perception of [ʃ]–[s] distinction: I. Temporal factors. Percept Psychophys 28:213–228
100. Martinet A (1955) Economie des Changements Phone tiques. Francke, Bern
101. Matthies M, Perrier P, Perkell JS, Zandipour M (2001) Variation in anticipatory coarticulation with changes in clarity and rate. J Speech Lang Hear Res 44:340–353
102. Mowrey R, Pagliuca W (1995) The reductive character of articulatory evolution. Rivista di Linguistica 7:37–124
103. Munson B, Beckman M, Edwards J (2012) Abstraction and specificity in early lexical representations: climbing the ladder of abstraction. In: Cohn A, Fougeron C, Huffman M (eds) The oxford handbook of laboratory phonology. Oxford University Press, Oxford, pp 288–309
104. Müller D (2013) Liquid dissimilation with a special regard to Latin. In: Sánchez Miret F, Recasens D (eds) Studies in phonetics, phonology and sound change in romance. Lincom, Munich, pp 95–109
105. Müller V, Harrington J, Kleber F, Reubold U (2011) Age-dependent differences in the neutralization of the intervocalic voicing contrast: evidence from an apparent-time study on East Franconian. Interspeech, Florence
106. Nielsen K (2011) Specificity and abstractness of VOT imitation. J Phon 39:132–142
107. Nielsen K (2014) Phonetic imitation by young children and its developmental changes. J Speech Lang Hear Res 57:2065–2075

108. Nittrouer S, Studdert-Kennedy M (1987) The role of coarticulatory effects in the perception of fricatives by children and adults. J Speech Hear Res 30:319–329
109. Nosofsky R (1988) Exemplar-based accounts of relations between classification, recognition, and typicality. J Exp Psychol Learn Mem Cogn 14:700–708
110. Öhman S (1966) Coarticulation in VCV utterances: spectrographic measurements. J Acoust Soc Am 39:151–168
111. Ohala J (1981) The listener as a source of sound change. In: Masek CS, Hendrick RA, Miller MF (eds) Papers from the parasession on language and behavior. Chicago Linguistic Society, Chicago, pp 178–203
112. Ohala J (1993) The phonetics of sound change. In: Jones C (ed) Historical linguistics: problems and perspectives. Longman, London, pp 237–278
113. Ohala J (2012) The listener as a source of sound change: an update. In: Solé M-J, Recasens D (eds) The initiation of sound change. perception, production, and social factors. John Benjamins, Amsterdam, pp 21–36
114. Ohala M, Ohala J (1992) Phonetic universals and Hindi segment duration. In: Proceedings of the international conference on spoken language processing. Edmonton, Alberta, pp 831–834
115. Osthoff H, Brugmann K (1878) Morphologische Untersuchungen auf dem Gebiete der indogermanischen Sprachen, Band I. Leipzig
116. Pardo J, Jay I, Krauss R (2010) Conversational role influences speech imitation. Atten Percept Psychophys 72:2254–2264
117. Pardo J, Gibbons R, Suppes A, Krauss R (2012) Phonetic convergence in college roommates. J Phon 40:190–197
118. Paul H (1886) Prinzipien der Sprachgeschichte, 2nd edn. Halle, Niemeyer
119. Phillips B (2006) Word frequency and lexical diffusion. Palgrave Macmillan, Basingstoke
120. Pierrehumbert J (2001) Exemplar dynamics: word frequency, lenition, and contrast. In: Bybee J, Hopper P (eds) Frequency effects and the emergence of lexical structure. John Benjamins, Amsterdam, pp 137–157
121. Pierrehumbert J (2003) Phonetic diversity, statistical learning, and acquisition of phonology. Lang Speech 46:115–154
122. Pierrehumbert J (2006) The next toolkit. J Phon 34:516–530
123. Pierrehumbert J, Beckman M, Ladd DR (2000) Conceptual foundations of phonology as a laboratory science. In: Burton-Roberts N, Carr P, Docherty G (eds) Phonological knowledge. Oxford University Press, Oxford, pp 273–303
124. Pierrehumbert J, Stonedahl F, Dalaud R (2014) A model of grassroots changes in linguistic systems. arXiv:1408.1985v1
125. Quené H (2013) Longitudinal trends in speech tempo: the case of Queen Beatrix. J Acoust Soc Am 133:452–457
126. Repp BH (1982) Phonetic trading relations and context effects: new experimental evidence for a speech mode of perception. Psychol Bull 92:81–110
127. Reubold U, Harrington J (In press) The influence of age on estimating sound change acoustically from longitudinal data. In: Wagner SE, Buchstaller I (eds) Panel studies of variation and change. Routledge Ltd, New York
128. Ruch H, Harrington J (2014) Synchronic and diachronic factors in the change from pre-aspiration to post-aspiration in Andalusian Spanish. J Phon 45:12–25
129. Sancier M, Fowler C (1997) Gestural drift in a bilingual speaker of Brazilian Portuguese and English. J Phon 25:421–436
130. Sankoff G, Blondeau H (2007) Language change across the lifespan: /r/ in Montreal Speech. Language 83:560–588
131. Schuchardt H (1885) Über die Lautgesetze: Gegen die Junggrammatiker. Oppenheim, Berlin
132. Sebanz N, Bekkering H, Knoblich G (2006) Joint action: bodies and minds moving together. Trends Cogn Sci 10:70–76
133. Shockley K, Richardson D, Dale R (2009) Conversation and coordinative structures. Top Cogn Sci 1:305–319

134. Siddins J, Harrington J, Reubold U, Kleber F (2014) Investigating the relationship between accentuation, vowel tensity and compensatory shortening. In: Proceedings of 7th speech prosody conference. Dublin, Ireland
135. Silverman D (2006) the diachrony of labiality in trique, and the functional relevance of gradience and variation. In: Goldstein L, Whalen D, Best C (eds) Papers in laboratory phonology 8. New Haven, Mouton de Gruyter, pp 133–154
136. Solé M (2007) Controlled and mechanical properites in speech: a review of the literature. In: Solé M-J, Beddor P, Ohala M (eds) Experimental approaches to phonology. Oxford University Press, Oxford, pp 302–321
137. Solé M (2009) Acoustic and aerodynamic factors in the interaction of features: the case of nasality and voicing. In: Vigário M, Frota S, Freitas MJ (eds) Phonetics and phonology: interactions and interrelations. John Benjamins, Amsterdam, pp 205–234
138. Solé M (2014) The perception of voice-initiating gestures. Lab Phonol 5:37–68
139. Stanford J, Kenny L (2013) Revisiting transmission and diffusion: an agent-based model of vowel chain shifts across large communities. Lang Var Chang 25:119–153
140. Stevens K (1972) The quantal nature of speech: evidence from articulatory-acoustic data. In: David E, Denes P (eds) Human communication: a unified view. McGraw-Hill, New York, pp 51–66
141. Stevens K (1989) On the quantal nature of speech. J Phon 17:3–45
142. Stevens M, Harrington J (2014) The individual and the actuation of sound change. Loquens 1(1):e003. doi:10.3989/loquens.2014.003
143. Stevens M, Reubold U (2014) Pre-aspiration, quantity, and sound change. Lab Phonol 5:455–488
144. Studdert-Kennedy M (1998) Introduction: the emergence of phonology. In: Hurford J, Studdert- Kennedy M, Knight C (eds) Approaches to the evolution of language. Cambridge University Press, Cambridge, pp 169–176
145. Stampe D (1969) The acquisition of phonetic representation. In: Papers from the fifth regional meeting of the chicago linguistic society, pp 443–454
146. Sweet H (1888) A history of English sounds. Clarendon Press, Oxford
147. Torreira F (2012) Investigating the nature of aspirated stops in Western Andalusian Spanish. J Int Phon Assoc 42:49–63
148. Trudgill P (1974) Linguistic change and diffusion: description and explanation in sociolinguistic dialect geography. Lang Soc 3:215–246
149. Trudgill P (1983) On dialect: social and geographical perspectives. New York University Press, New York
150. Trudgill P (2004) Dialect contact and new-dialect formation: the inevitability of colonial englishes. Edinburgh University Press, Edinburgh
151. Trudgill P (2008) Colonial dialect contact in the history of European languages: on the irrelevance of identity to new-dialect formation. Lang Soc 37:241–254
152. Trudgill P (2008) On the role of children, and the mechanical view: a rejoinder. Lang Soc 37:277–280
153. Vihman M (1980) Sound change and child language. In: Traugott E, La Brum R, Shepherd S (eds) Papers from the 4th international conference on historical linguistics. John Benjamins, Amsterdam, pp 303–320
154. Wang WS-Y (1969) Competing changes as a cause of residue. Language 45:9–25
155. Wardhaugh R, Fuller J (2015) An introduction to sociolinguistics, 7th edn. Wiley Blackwell, Chichester
156. Watson CI, Maclagan M, Harrington J (2000) Acoustic evidence for vowel change in New Zealand English. Lang Var Chang 12:51–68
157. Wedel A (2007) Feedback and regularity in the lexion. Phonology 24:147–185
158. Weinreich U, Labov W, Herzog M (1968) Empirical foundations for a theory of language change. In: Lehmann W, Malkiel Y (eds) Directions for historical linguistics. University of Texas Press, Austin, pp 95–195
159. Wells J (1982) Accents of English 2: the British Isles. Cambridge University Press, Cambridge

160. Wolfram W, Schilling-Estes N (2003) Dialectology and diffusion. In: Joseph B, Janda R (eds) The handbook of historical linguistics. Blackwell, Oxford, pp 713–735
161. Wright R (2003) Factors of lexical competition in vowel articulation. In: Local J, Ogden R, Temple R (eds) Papers in laboratory phonology VI. Cambridge University Press, Cambridge, pp 75–87
162. Yu A (2013) Individual differences in socio-cognitive processing and the actuation of sound change. In: Yu A (ed) Origins of sound change: approaches to phonologization. Oxford University Press, Oxford, pp 201–227
163. Yu A, Abrego-Collier C, Sonderegger M (2013) Phonetic imitation from an individual-difference perspective: Subjective attitude, personality, and 'autistic' traits. Plos One 8(9):e74746
164. Zellou G, Tamminga M (2014) Nasal coarticulation changes over time in Philadelphia English. J Phon 47:18–35

Chapter 7
Fostering User Engagement in Face-to-Face Human-Agent Interactions: A Survey

Chloé Clavel, Angelo Cafaro, Sabrina Campano
and Catherine Pelachaud

Abstract Embodied conversational agents are capable of carrying a face-to-face interaction with users. Their use is substantially increasing in numerous applications ranging from tutoring systems to ambient assisted living. In such applications, one of the main challenges is to keep the user engaged in the interaction with the agent. The present chapter provides an overview of the scientific issues underlying the engagement paradigm, including a review on methodologies for assessing user engagement in human-agent interaction. It presents three studies that have been conducted within the Greta/VIB platforms. These studies aimed at designing engaging agents using different interaction strategies (alignment and dynamical coupling) and the expression of interpersonal attitudes in multi-party interactions.

Keywords Embodied Conversational Agent · Interaction strategies · Socio-emotional behavior · User engagement

7.1 Introduction

One of the key challenges of human-agent interaction is to maintain user engagement. The design of engaging agents is paramount, whether in short-term human-agent interactions, or for building long-term relations between the user and the agent.

C. Clavel (✉) · A. Cafaro · C. Pelachaud
Telecom-ParisTech, LTCI, CNRS, Université Paris-Saclay, 46 rue Barrault, 75013 Paris, France
e-mail: chloe.clavel@telecom-paristech.fr

A. Cafaro
e-mail: angelo.cafaro@telecom-paristech.fr

C. Pelachaud
e-mail: catherine.pelachaud@telecom-paristech.fr
URL:http://www.tsi.telecom-paristech.fr/mm/en/themes-2/greta-team

S. Campano
Lab Object'ive, Object'ive, 96 98 rue de Montreuil, 75011 Paris, France
e-mail: scampano@object-ive.com

© Springer International Publishing Switzerland 2016
A. Esposito and L.C. Jain (eds.), *Toward Robotic Socially Believable Behaving Systems - Volume II*, Intelligent Systems Reference Library 106,
DOI 10.1007/978-3-319-31053-4_7

Many applications of human-agent interaction such as tutoring systems [45], ambient assisting living [12, 61] or virtual museum agents [24, 31, 58, 72] show the importance of the engagement paradigm. In ambient assisted living and tutoring systems, for example, the challenge is to maintain user engagement over many interactions, while in museum applications, the key issue is to invite visitors to interact and keep them engaged during the interaction for as long as possible.

In the human computer interaction literature, the issue of engagement is addressed from different angles. An interesting way of structuring such a literature is to rely on the distinction provided by Peters and colleagues [105], who distinguished the two following components underlying the engagement process: the *attentional involvement* and the *emotional involvement*. It is important to notice that, even though some studies focus more on one of these two components, they interleave, as the attention is driven by emotions. The definitions provided by Sidner and Dzikovska [125]— *"the process by which individuals in an interaction start, maintain and end their perceived connection to one another"*—focus on the *attentional involvement*, while [47, 63] focus on emotional engagement. In particular, in [63] they concentrate on empathic engagement—*"Empathic engagement is the fostering of emotional involvement intending to create a coherent cognitive and emotional experience which results in empathic relations between a user and a synthetic character"*. Another major distinction provided by Bickmore et al. [14] differentiates short-term versus long-term engagement. The former deals with user engagement in performing a task[1] while interacting with the agent. The latter implies much longer periods of interactions with the system and concerns the degree of involvement of the user over time.

The present chapter provides a review of the various literature dealing with the common objective of fostering user engagement in human-agent interactions (Sect. 7.2). The literature calls in different research fields ranging from social signal processing and affective computing to dialogue management and perceptive studies. Then, we focus on the description of examples of studies carried out inside a common platform—the Greta platform—(Sect. 7.3). Finally, we conclude and provide some tracks for the future design of engaging agents.

7.2   Designing Engaging Agents—State of the Art

"Embodied Conversational Agents (ECAs) are virtual anthropomorphic characters which are able to engage a user in real-time, multimodal dialogue, using speech, gesture, gaze, posture, intonation and other verbal and nonverbal channels to emulate the experience of human face-to-face interaction" [32]. Following this definition, in this section, we review the different research themes and issues involved in the usage of ECAs in order to foster user's engagement. These issues are represented in

---

[1]In this case, the task can just be interacting with the agent.

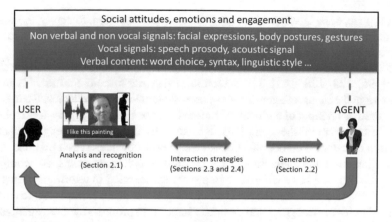

Fig. 7.1 The diagram shows the main research themes in the area of designing engaging agents. The focus is on the generation (*agent*) and the recognition (*user*) of socio-emotional behavior

the diagram in Fig. 7.1 which also shows the main multimodal channels adopted in human-agent interaction. non verbal and non vocal signals such as facial expressions, vocal signals such as prosody, and verbal content, such as word choice.

First, user's verbal and non verbal behavior should be taken into account by analysing the different modalities, on the one hand (Sect. 7.2.1) and the ECA should express relevant socio-emotional behavior through these modalities, on the other hand (Sect. 7.2.2). Then, multimodal dialogue management should be considered. In particular, we should address how to implement socio-emotional interactions between the user and the agent (Sects. 7.2.3 and 7.2.4). Finally, evaluation issues are paramount in human-agent interactions. The main questions are how to measure the impact of the design of engaging agent on user's impression (Sect. 7.2.5) and how to evaluate the 'success' of the interaction by analyzing the user engagement (Sect. 7.2.6).

7.2.1 Taking into account Socio-Emotional Behavior

ECAs aim at facilitating the socio-emotional interaction between the human and the machine. The socio-emotional aspect is a main prerequisite for a fluent interaction and thus for user's engagement. It relies on the development of agents endowed with socio-emotional abilities i.e. agents that are able to take into account user's social attitudes and emotions.

The user's expression of socio-emotional behavior can be both verbal and non-verbal (acoustic component of speech, facial expressions, gesture, body posture). Existing studies focused on the acoustic features such as prosody, voice quality or spectral features [27, 38, 124] and more generally on non-verbal features (posture, gesture, gaze, or facial expressions [89]) for emotion detection. Recently, the analysis of emotion has been integrated in a more general domain which considers also social behavior: the domain of social signal processing [138].

The analysis of non-verbal emotional content is widespread in human-agent interaction [96, 123, 128, 143]. The detection of the user's socio-emotional behaviors takes part in the inputs of agent's socio-emotional interaction strategies. It can also be considered as an input of a global user model for the building of long-term relations between the user and the agent [14]. It is strongly linked to user engagement: the detection of negative emotional states of the user is considered as a premise of user disengagement in the interaction [121]. Besides, avoiding (and thus, detecting) user frustration is also a key challenge to improve user learning in tutoring systems [77]. This statement is reinforced in [102] where they claim that it is important to consider students' achievement goals and emotions in order to promote their engagement and in [45], where they provide an *Affective auto-tutor agent* able to detect student's boredom and engagement. It is also interesting to note that some studies are not considering the socio-emotional level to detect engagement or disengagement but prefer to consider directly the signal level such as face location [17]. This last type of studies is efficient to detect user disengagement such as quitting the interaction. However, analyzing socio-emotional behavior as a cue of user engagement or disengagement supports the detection of subtler changes in the user engagement process.

The verbal content of emotions corresponds more to sentiment, opinion (see [87] for a discussion about the different terminologies) and attitude [120]. It begins to be integrated for the analysis of user socio-emotional behaviors. Natural language processing is yet not necessarily restricted to the analysis of the topic of user's utterance and can now give access to these socio-emotional cues [37]. In this way, in [142] they provided a system based on verbal cues that distinguished neutral, polite and frustrated user states. In [128], they proposed a classification between positive *vs.* negative user sentiment as an input of human-agent interaction. In [74] they provided a model of user's *attitudes* in verbal content grounded on the model described in [83] and dealing with the interaction context: as the previous agent's utterance can trigger or constrain an the user's expression of attitude, its target or its polarity, the model of the semantic and pragmatic features of agent's utterance is used to help the detection of user's attitude. Relying on the joint analysis of the agent's and the user's adjacent utterances, in [75] they provide a system able to detect user's like and dislike and devoted to the improvement of the social relationships between the agent and the user.

7.2.2 Generation of Agent's Socio-Emotional Behavior

An engaging agent in addition to perceiving the user's level of engagement should also be capable of maintaining it by exhibiting the appropriate socio-emotional behavior during the interaction. Two major issues arise regarding (1) the type of behavior to display and (2) when it should be exhibited. In this section we start by discussing the first issue continuing with the second one in Sect. 7.2.3. In particular, we describe several approaches adopted for the generation of multimodal behavior supporting the expression of social attitudes and emotions.

Expression of social attitudes As for modeling multimodal behaviors associated to social attitudes, several approaches rely on the Interpersonal Circumplex of attitudes proposed by Argyle [5] and on the correlation between specific behavior patterns and the expression of attitudes according to Burgoon and colleagues [22]. Two dimensions are considered, namely liking (or affiliation) and dominance (or status) [24, 36, 56, 76, 112]. Multimodal behaviors including gaze and head movement, body orientation, facial expression, use of personal space can be exhibited for expressing different attitudes. In order to find out and to be able to model such a correlation between behaviors and attitudes, different methodologies have been proposed. Bickmore and colleagues [16] have incorporated findings from social psychology to specify the behavior of their relational agent Laura. Experimental studies have been designed where human subjects were asked to indicate their perception of social attitudes of virtual agents [7, 8]. They show the importance of flirting tactics through gaze behaviors and expressive mimics in order to establish a first contact between the agent and the user [7] and that gaze behaviors and the linguistic expression of disagreeableness have a significant effect on the perception of dominance [8]. Several researchers collected and analyzed corpora of interacting human participants [36, 76], allowing the extraction of behaviors patterns linked to social attitudes. Ravenet et al. [112] used a crowd sourcing method asking human subjects to design agents with different attitudes by selecting multimodal behaviors through an interactive interface. The perception of a behavior (e.g. a smile) may vary depending on the chosen timing and context in which it is exhibited. For example, a smile followed by a gaze shift conveys a different attitude compared to a smile followed by a leaning toward the interlocutor. According to this, Chollet and colleagues [36] proposed to model the expression of an attitude as a sequence of behaviors. From the analysis of a corpus which has been annotated on two levels, attitudes and multimodal behaviors, sequences of multimodal behaviors linked to an attitude, are extracted using a sequence mining approach as suggested in [129].

While several studies focused on expressing social attitudes in face-to-face interaction, others have looked at the expression of an attitude by an agent in a multi-party interaction (e.g. in a conversing group) [55, 76, 112]. In such scenarios, the behavior exhibited by an agent needs to take into account the behaviors exhibited by other participants (for example in a group conversation) in order to compute the intended attitudes. In the project Demeanour [56], they modeled the posture and the gaze direction of virtual agents interacting with each other. The agents adapted their gaze direction, in particular their amount of mutual gaze, based on their attitudes toward each other. Ravenet et al. [114] went further in this direction and proposed a model of turn-taking. Depending on the attitudes toward each other (that may or may not be symmetrical), the agents adapt their spatial relations, their body orientation, gesture quality. They also change their manner to take the speaking turn or to handle interruption. For example, an agent that is dominant towards another has the tendency to interrupt the latter while it holds the speaking turn.

Expression of emotions Early ECAs models (c.f. [11, 103]) focused on the six prototypical expressions of emotions namely: anger, disgust, fear, joy, sadness and surprise (see [49]). However these expressions are barely used in interaction. Later on, researchers focused on endowing an agent with more subtle and more varied expressions of emotions. The proposed models can be distinguished by the theoretical model of emotions they used. Three main approaches can be reported: discrete emotion theory, dimension theory and appraisal theory. Computational models of the expressions of emotions rely on one of these models.

The discrete emotion theory introduced by Ekman [48] and Izard [68] claims that there is a small set of primary emotions that are universally produced and recognized. The expressions of these emotions can be blended as suggested by Ekman and Friesen [49]. As above mentioned, early models were built using findings from these theoretical models. Models of expression blending have been proposed in [21, 91]. Fuzzy logic was used to blend expressions of two emotions on the different parts of the face.

The dimensional theory describes emotions along a continuum over two [107, 117], or three [84], or even four dimensions [51]. The most common dimensions are pleasure, arousal and dominance. Emotions are no longer referred by a label (e.g., relief, regret) but by their coordinates in space. Computational models relying on dimensional representations make use of this continuum. They propose to create new expressions as the blending of facial expressions of known emotions placed in the 2D or 3D space. One of the first models was proposed by Ruttkay et al. [118]. The authors have developed a tool called Emotion Disc. Expressions of the six prototypical emotions are placed around a circle. The distance from the center to the outer of the circle indicates the intensity of the expression. A new expression can be created as a linear interpolation of these prototypical expressions. Other researchers [3, 136] proposed to calculate a new expression by interpolation between the closest known expressions. The interpolation can be done in 2D [136] and in 3D [3] space. Such approaches allow computing intermediate expressions from existing ones.

In [18, 59] they followed a very different approach. While the methods just presented are based on the prototypical expressions, these latter authors create a large set of facial expressions by composing randomly actions units defined with FACS [50]. Then they ask participants to rate the expressions along 2D [59] or 3D [18] space.

Appraisal theory views emotions as arising from the evaluation of an event, object, or person along different dimensions. In particular, the Componential Process Model (CPM) introduced by Courgeon et al. [119] makes predictions between how an event is appraised and the facial response. Few attempts [39, 97], have been made to implement how the facial responses are temporally organized to create the expression of emotion. Thus the expression of emotion does not correspond to a full-blown expression that arises in a block; rather it is made of a sequence of signals that arises and is composed on the face. In [92, 141], they pushed forward this idea. Based on a corpus analysis where multimodal behaviors have been annotated, the authors extract sequences of behaviors linked to emotions either manually [92] or automatically [141] using the T-pattern model developed by [81]. From the extraction of the data, Niewiadomski and colleagues [92] defined a set of rules that encompasses the spatial and temporal constraints of the signals in the sequences. Such models allow generating the expressions of emotions as sequences of temporally-ordered multimodal signals.

7.2.3 Socio-Emotional Interaction Strategies

In addition to the requirement of taking into account user socio-emotional behavior, on the one hand and to generate believable and engaging socio-emotional behavior of the agent, on the other hand, HCI requires to define the socio-emotional strategies linking the user input to the agent output. Existing strategies do not always have the explicit goal of fostering user engagement. In this paragraph, we focus on examples of strategies that have been explicitly used to foster user's engagement or to improve feelings of *rapport*, a concept which is strongly linked to engagement [14].

Providing backchannels and feedbacks is a key strategy for maintaining user engagement by providing agent's listening behaviors [73]. Thus, in the study of D'Mello and Graesser [45], the Auto-tutor agent provided feedback in order to help students to regulate their disengagement (boredom, etc.). In [122], the agent was able to generate multimodal backchannel (smile, nod, and verbal content) when it is listening to the user, and the timing of the backchannel—that is when to trigger the backchannel—was provided by probabilistic rules. In [135], they provided another rule-based model in order to predict when a backchannel has to be triggered as a reaction to prosody and pause behaviors. In [86] they used sequential probabilistic models, an interesting method to predict jointly when and how to generate backchannels in the listening phase of the agent. The timing issue of backchannel is close to the issue tackled in turn-taking strategies, that is, when the agent has to take or give the floor. As described in Sect. 7.2.5, researchers presented different turn-taking strategies and evaluated their role on the user's impressions [79, 80].

Politeness strategies are also associated with the concept of engagement. They provide to the agent a social intelligence [139] and allow it to be perceived as more engaged in the interaction [57]. In [4], politeness strategies were used as an answer to the expression of negative emotional states by the user to adjust the politeness level of their virtual guide. The more the interlocutor is in a negative emotional state, the more the guide has to be polite. However, Campano et al. [28] showed that in certain situations such as in video games, the agent has to express impoliteness to be more believable.

Endowing agents with humor may be a smart answer when the user is confused in front of some dysfunctions of the interaction system. Dybala and colleagues [47] proposed a humor-equipped casual conversational system (chatbot) and demonstrated that it enhances the user's positive engagement/involvement in the conversation.

A last example of smart strategies dedicated to improve user engagement is the management of agent's surprise. Bohus and Horvitz [17] proposed to communicate the robot's surprise when the user seemed to be disengaged in the interaction by using linguistic hesitation.

7.2.4 Alignment-Related Processes

Alignment [106] of ECA's behavior on the user is another strategy for improving user's engagement. Various approaches are used to design alignment processes or similar processes. These processes differ on the way they integrate the temporal and dynamic aspects. For example, mimicry is defined as the direct imitation of what the other participant produces [10], while synchrony is defined as the dynamic and reciprocal adaptation of temporal structures of behaviors between interactive partners [42]. The processes also differ on the levels at which they occur. At the lowest level, the processes concern the imitation of different modalities: body postures [34], gestures [85], accent and speech rate [54], phonetic realizations [98], word choice [53], repetitions [10, 140], syntax [19] and linguistic style [90].

The higher levels are mental, emotional or cognitive levels. Emotional resonance [60], affiliation [130]—that is alignment on users opinion or attitude (see Sect. 7.3.3)—and alignment at the level of concepts [20] are examples of high-level processes. But the different levels interleave. For example, copying gestures can be viewed as a way to establish and maintain an empathetic connection [33].

Alignment-related processes have been largely studied through linguistic studies dedicated to the observation of corpora. However, recent years have seen an increase of interest in the implementation of alignment-related processes in human-computer interactions, and in human-agent interactions, in particular. Implementations of alignment strategies in human-computer dialogues concerned mainly alignment on lexical and syntactic choices [23], while the human-agent *face-to-face* interactions further implementations of non verbal alignment. It is also interesting to notice that the terminology used in human-agent interaction is slightly different from the one used in corpus studies. In human-agent interaction, the terminology includes terms such as: mimicry [64], coordination [60, 71], synchrony [42], social/emotional resonance [60, 71], emotional mirroring [1], and dynamical coupling [109].

Some researchers attempted to implement complex alignment-related processes in simulated agent-agent interactions, dealing with social resonance and coverbal coordination in [71] and smile reinforcement between two virtual characters [95].

In summary, the literature on the design of socio-emotional interaction strategies is plentiful in the ECA community. Sophisticated interaction strategies such as alignment-related processes are even more frequent and begin to be effectively integrated in ECA platforms.

7.2.5 Impact on User's Impression

Users' evaluation of agents' behavior and interaction strategies are fundamental for designing believable and engaging agents. Recent studies focused on evaluating agents' nonverbal multimodal behavior during the first interaction and, some in particular, focused on the very initial moments. In light of Bickmore's distinction between short *vs.* long term engagement (cf. Sect. 7.1), evaluating the user's impressions of an agent addresses usability issues in both short and long term interactions. The idea is that a more engaging agent during the first interaction is likely to form a positive impression and be accepted by the user, thus promoting further interactions [9, 24].

There is a great deal of information that can be picked from observing an agent's multimodal behavior during the interaction, some of the relevant studies presented in this section mainly dealt with the user's impressions of the agent's friendliness, dominance, agreeableness, warmth and competence. Therefore, the emphasis has been on agent's characteristics such as interpersonal attitude towards the user, personality and skill level in a selected context (e.g. competence), that can be extrapolated from brief observations of multimodal behavior.

Maat et al. [79, 80] showed how a realization of a simple communicative function (for managing the interaction) could influence users' impressions of an agent. They focused on impressions of personality (agreeableness), emotion and social attitudes (i.e. friendliness) through different turn-taking strategies in human face-to-face conversations applied to their virtual agents in order to create different impressions of them. Fukayama and colleagues [52] proposed and evaluated a gaze movement model that enabled a virtual agent to convey different impression to users. They used an "eyes-only" agent on a black background and the impressions they focused on were affiliation (friendliness, warmth) and status (dominance, assurance). Similarly, Takashima et al. [132] evaluated the effects of different eye blinking rates of virtual agents on the viewers subjective impressions of friendliness (a blink rate of about 18 blink/min for the avatar makes a friendly impression), nervousness (higher blink rates reinforced nervous impressions) and intelligence (lower blink rates gave intelligent impression).

Niewiadomski and colleagues [93] analyzed how the emotional multimodal behavior of a virtual assistant expressing happiness, sadness and fear influenced user's judgments of agent's warmth, competence and believability. In particular, socially appropriate emotions expressed by the agent led to higher perceived believability. Then they also found that the perception of agents' believability was highly correlated to the two major socio-cognitive dimensions of warmth and competence.

In [25, 26] they investigated how users interpreted an agent's nonverbal greeting behavior (i.e. smile, gaze and proxemics) in a first encounter in terms of friendly interpersonal attitudes and extraverted personality [26]. In a follow-up study they discovered that a friendly interpersonal attitude expressed with more smiling and gazing at user is more relevant than expressing extraversion with proxemics behavior when it comes to decide whether to continue the interaction with an agent [25].

Bergmann et al. [9] studied how appearance and nonverbal behavior, in particular gestures, affected the perceived warmth and competence of virtual agents over time. Their goal was to study how warmth and competence ratings changed from a first impression after a few seconds to a second impression after a longer period of human-agent interaction, depending on manipulations of the virtual agent's appearance (robot-like character vs. anthropomorphic virtual agent) and gestural behavior (absent vs. present). Results indicated that impressions of warmth changed over time and depended on the agent's appearance. Evaluations of competence also changed but seemed to depend more on gestural behavior.

Virtual and robotic conversational agents have been deployed in public spaces for field studies. These deployments allowed researchers to move from the controlled laboratory settings to a more natural real life environment. Researchers have examined different engagement strategies in first user-agent encounters in such locations (e.g. museums, reception halls) where a multitude of users is present. Experiments conducted in these settings yielded more natural data, but they face a challenging environment that can be noisy, for example, in a museum there could be distracting or competing stimuli.

Gockley and colleagues [58] built a robot receptionist installed in a hall at Carnegie Mellon University, in USA. Valerie was able to give directions to visitors and look up the weather forecast while also exhibiting a compelling personality and character to encourage multiple visits over extended periods of time. The robot classified users according to an attentional zone based on their proximity and orientation (e.g. "engaged" visitors were close by the exhibit but not facing directly it).

Kopp et al. [72] installed Max in the Heinz Nixdorf Museums Forum (HNF), in Germany. Max was projected on a life-size screen and it was designed for being an enjoyable and cooperative interaction partner. It was able to engage with visitors in natural face-to-face conversations with a German voice accompanied by appropriate nonverbal behaviors such as facial expressions, gaze, and locomotion.

Cafaro et al. [24] conducted a study on Tinker at the Boston Museum of Science. Tinker was a human-sized conversational agent displayed as a cartoonish anthropomorphic robot that was capable of describing exhibits in the museum, giving directions, and discussing technical aspects of its own implementation. It used nonverbal conversational behavior, empathy, social dialogue, reciprocal self-disclosure and other relational behavior to establish social bonds with users. Tinker exhibited different greeting behaviors to approaching visitors (e.g. smiling behavior for friendliness). The visitors' commitment to interact with the agent was taken as behavioral measure of user's engagement. In the specific context of a first approach towards the exhibit, this measure was obtained by counting four possible actions from the moment the visitor was in the exhibit's area to the beginning of the interaction with the agent. These possible actions were: (i) walking past the exhibit, or (ii) finishing the approach towards the exhibit, (iii) following some instructions provided by Tinker on how to interact and (iv) effectively starting a conversation. There weren't significant differences among the groups receiving different greeting styles (i.e. no reaction, friendly and unfriendly), however trends seemed to indicate that the friendly version encouraged visitors to undertake more actions.

In summary, laboratory and field studies have been conducted to evaluate user's impressions of virtual and robotic agents. These studies focused on particular dimensions of first impressions such as interpersonal attitude and personality in order to make agents more engaging and accepted for long-term interactions.

7.2.6 Methodologies for Evaluating User Engagement in Human-Agent Interactions

So far we discussed state-of-the-art techniques and strategies for designing engaging ECAs in face-to-face or multi-party interactions with users. We also briefly reviewed some studies aimed at evaluating the impact of agents' exhibited socio-emotional behavior on user's impressions of ECAs. These studies focused on specific dimensions of user's impressions (e.g. agent's interpersonal attitude, competence, warmth) that are likely to improve the level of engagement with the agent. In this section, we move further from the mere assessment of users' first impressions by providing a brief survey on existing methodologies adopted by researchers to assess the user's engagement.

User engagement with an ECA, in general, can be measured via user self-reports (i.e. *subjective*); by monitoring the user's responses, tracking the user's body postures, intonations, head movements and facial expressions during the interaction (i.e. *objective*); or by manually logging behavioral responses of user experience (i.e. *behavioral* or *annotated*). This reflects a common categorization in experimental design [40]. A researcher could attempt to adopt any of the three above-mentioned approaches (or even combinations of those) to capture engagement.

Prior to providing some examples adopting these different methodologies, we should consider another factor that affects the assessment of engagement. In particular, the time window within which users should be asked (or measured) to express (or being detected) the level of engagement with an ECA. Reporting on paper or on a digital questionnaire sheet is the most popular approach to subjective assessment for user engagement, either asking the user upon showing a stimulus or at the end of a series of stimuli. However, there are two extremes that could be considered. One is focusing on real-time assessments *during* the interaction (mostly suitable, for example, when taking objective physiological measurements). On the other hand, there could be a longitudinal assessment taken over repeated interactions in a time span that might cover days, weeks or months of user-agent interactions. We refer to these different timings for assessing engagement as: *within-interaction* (during the interaction, for example at the end of the agent's turn), *end-interaction* and *over-several-interactions* (i.e. in longitudinal studies over multiple interactions).

Subjective assessments of engagement to date have been obtained through questionnaires with closed or open questions, or with structured interviews. Methods such as self-report closed questionnaires are constraining users into specific questionnaire items yielding data that can be easily used for analysis. However, there could be experimental noise in the responses. For example, participants might be biased after repeated interactions and there could be users' memory limitations about the perceived agent's behavior if asked at the end of interaction (*post-stimuli*), and self-deception (i.e. user not providing the true responses). Example of questionnaires adopted for measuring engagement can be found in an evaluation study presented in [67] and as a dimension of the Temple Presence Inventory (TPI) [78]. In [126], for instance, they adapted the TPI dimensions for studying user-robot engagement. Furthermore, in [46] they developed the Post-Lecture Engagement Questionnaire that required participants to self-report their engagement levels after each lecture. There were three questions which asked participants to rate their engagement at the beginning, middle, and end of each lecture. Participants indicated their ratings on a six-point scale ranging from (1) very bored to (6) very engaged.

Interviews may offer richer information, but the nature of these less structured data compared to quantitative data is harder to analyze. Example of these assessments are structured interviews or free text responses (leading to, for example, adjective analysis). Traum and colleagues [134] measured visitors' engagement when interacting with a pair of museum agents by adopting a mixed approach with both a self-report questionnaire and interviewing subjects at the end of the interaction.

The administration of subjective assessments is usually done at the end of the interaction or over several interactions. It might be intrusive and hard to obtain within-interaction (i.e. questions appearing during the interaction).

Finally, an example of longitudinal assessment with focus on building a working alliance between user and agent in the health domain can be found in [13].

Objective studies rely on automatic detection of physiological [35], verbal or non verbal signals that can be linked to engagement. They can be driven both within the interaction, at the end of the interaction or longitudinally (over several interactions). Analysis of user engagement within the interaction can be provided by automatic analysis such as the one described in Sect. 7.2.1 by analyzing speech prosody or body postures, emotions and attitudes in order to infer user engagement. Unlike subjective self-reports, automatic analysis provides both information on the evolution of the user engagement *within* the interaction and global evaluation of user engagement. Simple automatic measurements at the interaction level can also be provided: Bickmore and colleagues [15] measured the total time in minutes each visitor spent with a relational agent installed in a museum.

Another way to assess user engagement within the interaction and to capture its evolution along the interaction is to carry out *behavioral studies*. Sidner et al. [126], thus provided annotations of videotaped interactions between the user and a robot including the duration of the interaction, the amount of shared looking (looking the same object), mutual gazes (looking at each other), looking at the robot during the humans turn and overall amount of time the user spent looking at the robot. In [88] they describe a study in which a user interacted with an ECA and an external observer watched this interaction. In addition, a push-button device was given to both the user and the observer. The user was instructed to press the button when the agents explanation was boring and the user would like to change the topic. The observer was instructed to press the button when the user looked bored and distracted from the conversation.

A strength of behavioral and objective studies is their lack of intrusion into the user-agent interaction experience. However, objective studies can be exposed to detection errors, for example when automatically recognizing user's multimodal behavior and behavioral studies are subjected to labelers subjectivity even though they are more shielded from subject bias than subjective studies.

Like subjective studies, objective and behavioral studies are relevant for a longitudinal assessment of engagement over several interactions, an issue which is especially important for applications such as assisted living [14].

7.2.7 Summary of the Key Points for the Design of Engaging Agent

The socio-emotional component has a key role in the design of engaging agent. The literature on users' emotion recognition and on the generation of the agent's emotional behavior begins to have a quite long tradition and to offer a range of satisfactory tools for non-verbal aspects. Further work needs to be done concerning the analysis of the user's verbal socio-emotional content and the use of user socio-emotional behavior in socio-emotional interaction strategies. Besides, the integration of social component with the generation of agents' social stances is more recent and is a promising contribution to the engagement paradigm in human-agent interaction. Next section provides a summary of studies dealing with these three scientific challenges: integration of the verbal content (Sect. 7.3.3) and of the non verbal content in socio-emotional interaction strategies (Sect. 7.3.2); and the expression of social stances in multiparty group interaction (Sect. 7.3.4).

7.3 Overview of Studies Carried Out in GRETA and VIB

The design of engaging agents has been implemented by several studies around a same platform that makes it possible to integrate the different modules required for an engaging human-agent interaction—from the detection of user socio-emotional behavior to the generation of agent socio-emotional behaviors: the Greta system and VIB platform. In this section, we first present the architecture of the Greta system, then we show the extension of this system in the VIB platform, and finally, we present three different studies dedicated to foster user engagement that have been implemented in VIB/Greta. The first two studies deal with computational models of alignment-related processes (dynamical coupling and alignment) as described in Sect. 7.2.4. In particular, Sect. 7.3.2 shows how dynamical coupling can improve user experience and contribute to user engagement and Sect. 7.3.3 focuses on alignment strategies and their impact on user engagement. The third study focuses on user experience and engagement in multiparty interactions with conversing group of virtual agents.

7.3.1 Greta System and VIB Platform

The Greta system allows a virtual or physical (e.g. robotic) embodied conversational agent to communicate with a human user [94, 95]. The global architecture of the system is depicted in Fig. 7.2. It is a SAIBA compliant architecture (SAIBA is a common framework for the autonomous generation of multimodal communicative behavior in Embodied conversational agents [70]). The main three components are: (1) an *Intent Planner* that produces the communicative intentions and handles the emotional state of the agent; (2) a *Behavior Planner* that transforms the communicative intentions received in input into multimodal signals and (3) a *Behavior Realizer* that produces the movements and rotations for the joints of the ECA.

A *Behavior Lexicon* (i.e. Agent Behavior Specification in Fig. 7.2) contains pairs of mappings from communicative intentions to multimodal signals. The Behavior Realizer instantiates the multimodal behaviors, it handles the synchronization with speech and generates the animations for the ECA.

The information exchanged by these components is encoded in specific representation languages defined by SAIBA. The representation of communicative intents is done with the Function Markup Language (FML) [65]. FML describes communicative and expressive functions without any reference to physical behavior, representing in essence what the agent's mind decides. It is meant to provide a semantic descrip-

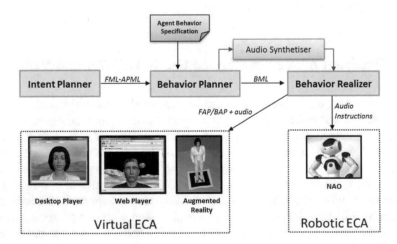

Fig. 7.2 The Greta system

tion that accounts for the aspects that are relevant and influential in the planning of verbal and nonverbal behavior. Greta uses an FML specification named *FML-APML* and based on the Affective Presentation Markup Language (APML) introduced by [41]. FML-APML tags encode the communicative intentions following the taxonomy defined in [108], where a communicative function corresponds to a pair (*meaning,signal*). The meaning element is the communicative intent that the ECA aims to accomplish, whereas the signal element indicates the multimodal behavior exhibited in order to achieve the desired communicative intent.

The multimodal behaviors to express a given communicative function to achieve (e.g. facial expressions, gestures and postures) are described by the Behavior Markup Language (BML) [70, 137].

The Greta system has been embedded in the Virtual Interactive Behavior (VIB) platform [99]. An overview of the VIB architecture is shown in Fig. 7.3. VIB enhances Greta with additional components that allow the ECA to detect its environment (i.e. *Perceptive Space* in Fig. 7.3), and to interact with the user while constantly updating the agent's mental and emotional states. Thus, an ECA's mental state includes information such as beliefs, goals, emotions and social attitudes.

The agent's emotional state is computed with the FaTiMa emotion model by [43]. A dialogue manger computes the utterances spoken by the agent as a function of both its mental state and previous verbal content exchanged with the user. Currently VIB integrates the DISCO dialogue manager developed by [116]. The output of this component is sent to the agent's intent planner.

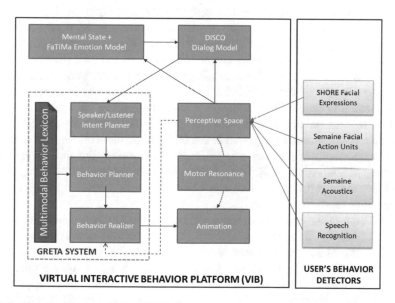

Fig. 7.3 Global architecture of the VIB platform

Different external tools plugged-in the VIB platform (i.e. SHORE facial expressions, SEMAINE facial action units and acoustics, and speech recognition as shown on the right side of Fig. 7.3) allow an agent to detect and interpret user's audio-visual input cues captured with devices such as cameras, Microsoft's Kinect and microphones. These information are provided to the agent via the *Perceptive Space* module. A direct link between this module and the *Behavior Realizer* allows the agent to exhibit reactive behaviors by quickly producing the behavior to exhibit in response to user's behavior, as for back-channels for example.

Finally, the *Motor resonance* manages the direct influence of the socio-emotional behaviors of the user (agent perceptive space) to the ones of the agent (agent production space) without cognitive reasoning. In particular, it allows the ECA to dynamically mimic the behavior of the user.

7.3.2   Modeling Dynamical Coupling

The study presented in this section focused on the *Motor Resonance* module in the GRETA platform. This study is about mirroring of human laughter by an ECA during an interaction. We refer to this process as *dynamical coupling*. The tool supporting the modeling of dynamic coupling in the platform can be used for other communicative functions. An interface allows us to connect the detected users inputs to ECA's animation parameters through a neural network.

Laughter is a social signal that has many functions in dialogue. For example, it allows someone to display a feeling of pleasure consecutively to positive events, such as receiving compliments [110] or perceiving a humorous stimulus. Laughter also serves to hide one's embarrassment [66], or to be cynical. It helps to create social bonds within groups [2], and regulates the speech flow in conversation [110]. These socio-emotional communicative functions are important in interaction. It is then important to enable ECAs to laugh in order to improve the quality of human-agent interaction, and to enhance user involvement. To this end, we defined a model of laughter [44], currently integrated in the GRETA architecture, and we conducted an evaluation that explores the role of laughter mirroring (dynamic coupling) in human-agent interaction [100]. Our goal was to study how the adaptive capabilities of an ECA, through the imitation of user's behaviors, could enhance user experience during human-agent interaction.

The setting of the experiment was an interactive installation called LoL, Laugh out Loud (ref). In this setting, a user and an ECA are listening to music inspired from the compositions by P. Shickele P.D.Q Bach. These recordings were created with the aim of making the listeners laugh. The ECA is able to tune on the fly the behavioral expressivity if its laughter according to the user's behavioral expressivity, hence creating a phenomenon of dynamic coupling between the ECA and the user. The parameters for the expressivity of ECA's laughter are the torso orientation, and the amplitude of laughter movements. For example, if the user does not laugh and or does not move at all, ECA's laughter behavior will be inhibited. On the contrary, if the user laughs out loud and moves a lot, ECA's laughter will be amplified.

The experimental study was conducted with 32 participants. Two conditions were tested: (i) the ECA takes user's behaviors into account to modulate the expressivity of its laughter, (ii) the ECA does not take user's behaviors into account. Once the participants had listened to two short musical compositions, they had to answer a questionnaire measuring ECA's social presence. The analysis of the results revealed that when the ECA takes user's behavior into account to modulate its laughter, its social presence as perceived by the participants is greater than when it does not. The participants had the feeling that it was easier to interact with the ECA, and they had the impression they were both in the same place and that they laughed together.

In this study, the ECA behavior was generated by taking into account the user's behavior. This modulation, which takes into account human acoustic and movement features, acts upon several parameters controlling agent's animation. We chose to inhibit or to amplify the intensity of ECA's behaviors in mirroring the intensity of user's behaviors. Mirroring can be seen as a form of *alignment* between an ECA and a user. The next section presents a study exploring verbal alignment between an ECA and a user.

7.3.3 Enhancing User Engagement through Verbal Alignment by the Agent

A model of verbal alignment allowing an ECA to share appreciations with a user [30], referred as the *Appreciation Module*, was integrated in the GRETA platform as an Intention Planner. It takes as inputs ECA's preferences encoded in the *Agent Mind*. The *Appreciation Module* provides supplementary functionalities for dialogue management, in addition to the DISCO dialogue manager [116], integrated in the GRETA platform.

The development of the *Appreciation Module* is conducted in the framework of the national French project A1:1. The project's goal is to set a life-sized ECA in a museum where it plays the role of a visitor. The module, in particular, aims at enabling the ECA to *engage* museum visitors by sharing appreciations on different topics, such as an artwork, or a specific painting style. Expressing evaluation opinion or judgment is a basic activity for visitors in a museum [127], and it is important to build *rapport* and *affiliation* between two speakers, which contributes to their *engagement* [144]. Our model is twofold: it focuses on *how* an ECA can generate appreciation sentences, and *when* the ECA should effectively use them.

We modeled two types of *alignment processes* occurring during the sharing of appreciations between an user and the ECA: alignment at the lexical level through *other-repetition* (OR), and alignment at the level of *polarity* between a user's appreciation and an ECA's appreciation on the same topic. OR is the intentional repetition

Fig. 7.4 A user interacting
with Leonard during our
study conducted in the
laboratory

by the hearer of part of what the speaker has just said, in order to convey a communicative function that was not present in the first instance [6, 10, 104, 133], such as
an emotional stance [131].

Our computational model enables an ECA to express emotional stances with other-
repetitions [30]. This model is grounded on a previous analysis on the SEMAINE
corpus [29], where we found several occurrences of ORs expressing emotional
stances. Our model integrates 3 *emotional stances*: surprise, positive appreciation,
negative appreciation. The selected emotional stance depends on user's appreciation
and ECA's preferences, and they are expressed by the ECA in the form of a verbal
appreciation, as defined in [82] (e.g.: "I consider it beautiful.").

An evaluation of the appreciation sentences generated by the model was conducted
with the ECA Leonard designed for the A1:1 project. We simulated a small museum
in our laboratory; We hung 4 pictures of existing artworks in the corridor. Each
participant was asked to watch them, then to talk to Leonard (see Fig. 7.4), and
ultimately fill in a questionnaire.

Thirty-four participants took part in the experiment. The results issued from subjective reports showed that the presence or absence of ORs in ECA's appreciations
does not seem to have an effect on the perception of user's own engagement, or on
ECA's believability perceived by the user. However, the presence of ORs in ECA's
appreciations had a positive effect on participants' feeling that they shared the same
appreciations as the ECA.

To improve these results, we developed an extension of the previous model dedicated at deciding *when* to trigger an exchange of appreciations between the ECA and
the user [31]. This sharing of appreciations, represented as a task, is added on the fly
in the dialogue plan when the user shows a low level of engagement while interacting
with an ECA. For future work, we plan to conduct an evaluation of the model with
different conversational strategies, such as triggering a sharing of appreciations when
user engagement is high versus low.

7.3.4   Engaging Users in Multiparty Group Interaction with the Expression of Interpersonal Attitudes

Simulating group interactions and expressing social attitudes among participants can be hard to achieve. The expression of the agent's interpersonal attitude with multimodal behavior in user-agent face-to-face interactions is supported by the Greta platform [111]. However, moving to a more complex multi-party group interaction required a more powerful framework that integrated the Greta platform with the Impulsion AI Engine [101]. This latter engine combined a number of reactive social behaviors, including those reflecting Hall's personal space theory [62] and Kendon's F-formation systems [69], in a general steering framework inspired by Reynolds and colleagues [115]. The engine supports the generation of agents' reactive behaviors to make them aware of the user's presence (for example in avatar based interactions), so that users feel engaged in interaction with an agent or a group.

Impulsion's complete management of position and orientation was used in conjunction with Greta's behavior planner for facial expressions and gestures generation, in order to produce believable and dynamic group formations expressing different interpersonal attitudes both among other members of a group of agents, thus named *In-group* attitude, and towards the user, therefore named *out group* attitude [113]

Interpersonal attitudes shape how one person relates to another person [5, p.85], in particular affiliation, according to Argyle's status and affiliation model [5, p.86], indicates the degree of liking or friendliness, ranging from unfriendly to friendly, towards another person. In the context of engagement, expressing high affiliation (i.e. friendliness) represents a valuable means for showing interest into interacting with one person.

In the context of user-agents interaction within a 3D serious game environment, the effects of both *in* and *out-group* attitude (affiliation dimension) on user's presence evaluations of a group of four agents and user's proxemics behavior in the 3D environment were studied. In two separate trials subjects had to complete the task of (1) joining a group of four agents, composed by two males and two females, and (2) reaching a point behind the group of agents with their own avatar in third person view (Fig. 7.5).

The different levels of attitude were obtained by exhibiting, for example in a friendly out-group case, smiling behavior, gazing more at the user (compared to the unfriendly case) and opening (i.e. making physical space) when the user's avatar was within the social distance of the group, according to Hall's areas [62]. The in-group attitude levels were obtained by changes of voice volume, gestures amplitude and speed, proximity among the agents, number of gaze at others, smiling behavior and turn duration.

In conclusion, results indicated that expressing interpersonal attitudes in multi-party group interaction had an impact on the evaluation of agent's presence assessed by users when those attitudes were expressed towards the user (out-group) regardless of the attitude expressed among the agents (in-group). The social presence (including engagement level) of a group of agents is dramatically reduced when an unfriendly

Fig. 7.5 The image shows a screen shot of the 3D environment as seen by the user in third person view with the avatar walking towards the group of agents in one of the conditions within the study on interpersonal attitudes in multi-party group interaction

attitude is expressed towards users. Interestingly, users in the first task (i.e. join a group) chose to get closer to those groups having unfriendly out-group attitude, possibly due to the lack of openness exhibited by the group, thus users pushed more their avatar in order to obtain a reaction. Whereas in the second task (i.e. reach destination behind the group) users walked through those having both in and out-group unfriendly attitude. This was possibly due to the bigger interpersonal space among the agents.

7.4 Conclusion and Perspectives

Considering engagement in human-agent interactions is a promising way for addressing the scientific challenges involved in generating fluent social interactions between the users and the agents. Research has unraveled many aspects concerning the detection and the generation of emotional behaviors but considering social attitudes is still an emerging topic. Existing socio-emotional interaction strategies pay more and more attention to fostering user engagement not only within a single interaction but also over several interactions. The success of socio-emotional interaction strategies can be thus evaluated by focusing on user engagement, and the present chapter provided a view of different methodologies that are used for this evaluation, from subjective tests to automatic measurements.

The work carried out around the Greta/VIB platform takes a step in this direction by providing subjective assessments of user engagement aiming to evaluate the interaction strategies (alignment and dynamic coupling) and the expression of interpersonal attitudes in multi-party interactions. Current work on this platform concerns the integration of a system able to detect user's likes and dislikes; the development of further interaction strategies such as politeness strategies and the management of turn-taking between the user and the agent. We hope such integration will contribute to provide more fluent interactions and improve the user's engagement.

Acknowledgments The authors would like to thank the GRETA team for its contributions to the Greta and Vib platforms. This work has been supported by the French collaborative project A1:1, the european project ARIA-VALUSPA, and performed within the Labex SMART (ANR-11-LABX-65) supported by French state funds managed by the ANR within the Investissements d'Avenir programme under reference ANR-11-IDEX-0004-02.

References

1. Acosta JC, Ward NG (2011) Achieving rapport with turn-by-turn, user-responsive emotional coloring. Speech Commun 53(910):1137–1148
2. Adelswärd V (1989) Laughter and dialogue: the social significance of laughter in institutional discourse. Nordic J Linguist 12(02):107–136
3. Albrecht I, Schröder M, Haber J, Seidel H (2005) Mixed feelings: expression of non-basic emotions in a muscle-based talking head. Virtual Real 8(4):201–212
4. Andre E, Rehm M, Minker W, Bühler D (2004) Endowing spoken language dialogue systems with emotional intelligence. Affective dialogue systems. Springer, Berlin, pp 178–187
5. Argyle M (1988) Bodily communication, 2nd edn. Methuen, New York
6. Bazzanella C (2011) Redundancy, repetition, and intensity in discourse. Lang Sci 33(2):243–254
7. Boo N, André E, Tober S (2009) Breaking the ice in human-agent communication: eye-gaze based initiation of contact with an embodied conversational agent. In: Proceedings of 9th international conference on Intelligent virtual agents, IVA, Amsterdam, The Netherlands, Sept 14–16, 2009, pp 229–242
8. Bee N, Pollock C, André E, Walker M (2010) Bossy or wimpy: expressing social dominance by combining gaze and linguistic behaviors. In: Proceedings of the 10th international conference on intelligent virtual agents, Springer, Berlin, Heidelberg, IVA'10, pp 265–271
9. Bergmann K, Eyssel F, Kopp S (2012) A second chance to make a first impression? how appearance and nonverbal behavior affect perceived warmth and competence of virtual agents over time. In: Nakano Y, Neff M, Paiva A, Walker M (eds) Intelligent virtual agents, Lecture Notes in Computer Science, vol 7502, Springer, Berlin Heidelberg, pp 126–138, doi:10.1007/978-3-642-33197-8_13
10. Bertrand R, Ferré G, Guardiola M et al (2013) French face-to-face interaction: repetition as a multimodal resource. Interaction, Coverbal Synchrony in Human-Machine, p 141
11. Beskow J (1997) Animation of talking agents. In: Benoit C, Campbell R (eds) Proceedings of the ESCA workshop on audio-visual speech processing. Rhodes, Greece, pp 149–152
12. Bickmore T, Giorgino T (2006) Health dialog systems for patients and consumers. J biomed inf 39(5):556–571
13. Bickmore T, Gruber A, Picard R (2005) Establishing the computer patient working alliance in automated health behavior change interventions. Patient Educ Couns 59(1):21–30. doi:10.1016/j.pec.2004.09.008
14. Bickmore T, Schulman D, Yin L (2010) Maintaining engagement in long-term interventions with relational agents. Appl Artif Intell 24(6):648–666
15. Bickmore T, Pfeifer L, Schulman D (2011) Relational agents improve engagement and learning in science museum visitors. Intelligent virtual agents. Springer, Berlin, pp 55–67
16. Bickmore TW, Picard RW (2005) Establishing and maintaining long-term human-computer relationships. ACM Trans Comput-Hum Interact 12(2):293–327
17. Bohus D, Horvitz E (2014) Managing human-robot engagement with forecasts and... um... hesitations. In: Proceedings of the 16th international conference on multimodal interaction, ACM, pp 2–9
18. Boukricha H, Wachsmuth I, Hofstätter A, Grammer K (2009) Pleasure-arousal-dominance driven facial expression simulation. In: Proceedings of third international conference and

workshops on Affective computing and intelligent interaction, ACII, Amsterdam, The Netherlands, Sept 10–12 2009, pp 1–7

19. Branigan HP, Pickering MJ, Cleland AA (2000) Syntactic co-ordination in dialogue. Cognition 75(2):B13–B25

20. Brennan SE, Clark HH (1996) Conceptual pacts and lexical choice in conversation. J Exp Psychol: Learn, Mem, Cognit 22(6):1482

21. Bui TD, Heylen D, Poel M, Nijholt A (2001) Generation of facial expressions from emotion using a fuzzy rule based system. In: Stumptner M, Corbett D, Brooks M (eds) Proceedings of 14th Australian joint conference on artificial intelligence (AI 2001). Springer, Adelaide, Australia, pp 83–94

22. Burgoon J, Buller D, Hale J, de Turck M (1984) Relational messages associated with nonverbal behaviors. Hum Commun Res 10(3):351–378

23. Buschmeier H, Bergmann K, Kopp S (2009) An alignment-capable microplanner for natural language generation. In: Proceedings of the 12th European workshop on natural language generation, association for computational linguistics, pp 82–89

24. Cafaro A, Vilhjálmsson HH (2015) First impressions in human-agent virtual encounters. ACM Trans Comput-Hum Interact (TOCHI, forthcoming)

25. Cafaro A, Vilhjálmsson HH, Bickmore TW, Heylen D, Schulman D (2013) First impressions in user-agent encounters: the impact of an agent's nonverbal behavior on users' relational decisions. In: Proceedings of the 2013 international conference on autonomous agents and multi-agent systems, international foundation for autonomous agents and multiagent systems, Richland, SC, AAMAS '13, pp 1201–1202. http://dl.acm.org/citation.cfm?id=2484920.2485142

26. Cafaro A, Vilhjlmsson H, Bickmore T, Heylen D, Jhannsdttir K, Valgarsson G (2012) First impressions: users judgments of virtual agents personality and interpersonal attitude in first encounters. In: Nakano Y, Neff M, Paiva A, Walker M (eds) Intelligent virtual agents, Lecture notes in computer science, vol 7502, Springer, Berlin Heidelberg, Richland, SC, pp 67–80. doi:10.1007/978-3-642-33197-8_7

27. Callejas Z, Griol D (2011) López-Cózar R (2011) Predicting user mental states in spoken dialogue systems. EURASIP J Adv Signal Proc 1:6

28. Campano S, Sabouret N (2009) A socio-emotional model of impoliteness for non-player characters. In: 3rd international conference on affective computing and intelligent interaction and workshops. ACII 2009. IEEE, pp 1–7

29. Campano S, Durand J, Clavel C (2014) Comparative analysis of verbal alignment in human-human and human-agent interactions. In: Proceedings of the 9th international conference on language resources and evaluation (LREC'14), European language resources association (ELRA)

30. Campano S, Langlet C, Glas N, Clavel C, Pelachaud C (2015) An eca expressing appreciations. In: First international workshop on engagement in human computer interaction ENHANCE'15 held in conjuction with the 6th international conference on affective computing and intelligent interaction (ACII 2015), IEEE, sept 2015 (to appear)

31. Campano S, Clavel C, Pelachaud C (2015) I like this painting too: when an eca shares appreciations to engage users. In: Proceedings of the 14th international joint conference on Autonomous agents and multiagent systems

32. Cassell J (2000) Embodied conversational agents. MIT press, Cambridge

33. Castellano G, Mancini M, Peters C, McOwan PW (2012) Expressive copying behavior for social agents: a perceptual analysis. IEEE Trans Syst, Man Cybern, Part A: Syst Hum 42(3):776–783

34. Chartrand TL, Bargh JA (1999) The chameleon effect: the perception-behavior link and social interaction. J pers soc psychol 76(6):893

35. Choi A, Melo CD, Woo W, Gratch J (2012) Affective engagement to emotional facial expressions of embodied social agents in a decision-making game. Comput Animat Virtual Worlds 23(3–4):331–342

36. Chollet M, Ochs M, Clavel C, Pelachaud C, (2013) A multimodal corpus approach to the design of virtual recruiters. In, (2013) humaine association conference on affective computing and intelligent interaction, ACII 2013. Geneva, Switzerland, Sept 2–5:19–24
37. Clavel C, Callejas Z (2015) Sentiment analysis: from opinion mining to human-agent inter-action. IEEE Trans Affect Comput 99:1. doi:10.1109/TAFFC.2015.2444846
38. Clavel C, Vasilescu I, Devillers L, Richard G, Ehrette T (2008) Fear-type emotions recognition for future audio-based surveillance systems. Speech Commun 50:487–503
39. Courgeon M, Clavel C, Martin J (2014) Modeling facial signs of appraisal during interaction: impact on users' perception and behavior. In: International conference on autonomous agents and multi-agent systems, AAMAS '14, Paris, France, May 5–9 2014, pp 765–772
40. Creswell JW (2013) Research design: Qualitative, quantitative, and mixed methods approaches (Sage publications)
41. De Carolis B, Pelachaud C, Poggi I, Steedman M (2004) Apml, a markup language for believ-able behavior generation. In: Prendinger H, Ishizuka M (eds) Life-like characters, cognitive technologies, Springer, Berlin Heidelberg, pp 65–85. doi:10.1007/978-3-662-08373-4_4
42. Delaherche F, Chetouani M, Mahdhaoui A, Saint-Georges C, Viaux S, Cohen D (2012) Interpersonal synchrony: a survey of evaluation methods across disciplines. IEEE Trans Affect Comput 3(3):349–365
43. Dias J, Paiva A (2005) Feeling and reasoning: a computational model for emotional agents. In: Proceedings of 12th Portuguese conference on artificial intelligence, EPIA 2005, Springer, pp 127–140
44. Ding Y, Prepin K, Huang J, Pelachaud C, Artières T (2014) Laughter animation synthesis. In: Proceedings of the 2014 international conference on autonomous agents and multi-agent systems, AAMAS '14, pp 773–780
45. D'Mello S, Graesser A (2013) AutoTutor and affective autotutor: learning by talking with cognitively and emotionally intelligent computers that talk back. ACM Trans Interact Intell Syst 2(4):1–39
46. D'Mello S, Olney A, Williams C, Hays P (2012) Gaze tutor: a gaze-reactive intelligent tutoring system. Int J hum-comput stud 70(5):377–398
47. Dybala P, Ptaszynski M, Rzepka R, Araki K (2009) Activating humans with humor - a dialogue system that users want to interact with. IEICE Trans Inf Syst E92-D(12):2394–2401
48. Ekman P (2003) Emotions revealed. Times Books (US), London, New York. Weidenfeld and Nicolson (world)
49. Ekman P, Friesen W (1975) Unmasking the face: a guide to recognizing emotions from facial clues. Prentice-Hall Inc, Englewood Cliffs
50. Ekman P, Friesen W, Hager J (2002) The facial action coding system, 2nd edn. Research Nexus eBook, London, Salt Lake City, Weidenfeld and Nicolson (world)
51. Fontaine JR, Scherer KR, Roesch EB, Ellsworth P (2007) The world of emotion is not two-dimensional. Psychol Sci 13:1050–1057
52. Fukayama A, Ohno T, Mukawa N, Sawaki M, Hagita N (2002) Messages embedded in gaze of interface agents–impression management with agent's gaze. In: Proceedings of the SIGCHI conference on human factors in computing systems, ACM, New York, NY, USA, CHI '02, pp 41–48. doi:10.1145/503376.503385
53. Garrod S, Anderson A (1987) Saying what you mean in dialogue: a study in conceptual and semantic co-ordination. Cognition 27(2):181–218
54. Giles H, Coupland N, Coupland J (1991) Contexts of accommodation: developments in applied sociolinguistics, Cambridge, Cambridge University Press, chap 1–Accommodation theory: communication context, and consequence
55. Gillies M, Ballin D (2003) A model of interpersonal attitude and posture generation. In: Proceedings of 4th international workshop on Intelligent agents, IVA 2003, Kloster Irsee, Germany, Sept 15–17 2003, pp 88–92
56. Gillies M, Ballin D (2004) Integrating autonomous behavior and user control for believable agents. In: 3rd International joint conference on autonomous agents and multiagent systems AAMAS, New York, NY, USA, August 19–23 2004, pp 336–343

57. Glas N, Pelachaud C (2014) Politeness versus perceived engagement: an experimental study. In: Proceedings of the 11th international workshop on natural language processing and cognitive science

58. Gockley R, Bruce A, Forlizzi J, Michalowski M, Mundell A, Rosenthal S, Sellner B, Simmons R, Snipes K, Schultz A, Wang J (2005) Designing robots for long-term social interaction. In: IEEE/RSJ International conference on intelligent robots and systems (IROS), pp 1338–1343. doi:10.1109/IROS.2005.1545303

59. Grammer K, Oberzaucher E (2006) The reconstruction of facial expressions in embodied systems. ZiF: mitteilungen, vol 2

60. Gratch J, Kang SH, Wang N (2013) Using social agents to explore theories of rapport and emotional resonance. Oxford University Press, Oxford, Social emotions in nature and artifact, p 181

61. Griol D, Molina JM, Callejas Z (2014) Modeling the user state for context-aware spoken interaction in ambient assisted living. Appl Intell 40:1–23

62. Hall ET (1966) The hidden dimension. Doubleday, Garden City

63. Hall L, Woods S, Aylett R, Newall L, Paiva A (2005) Achieving empathic engagement through affective interaction with synthetic characters. Affective computing and intelligent interaction. Springer, Heidelberg, pp 731–738

64. Hess U, Philippot P, Blairy S (1999) The social context of nonverbal behavior mimicry. cambridge University Press, Cambridge, p 213

65. Heylen D, Kopp S, Marsella SC, Pelachaud C, Vilhjálmsson HH (2008) The next step towards a function markup language. Proceedings of the 8th international conference on Intelligent Virtual Agents. Springer, Berlin, IVA, pp 270–280

66. Huber T, Ruch W (2007) Laughter as a uniform category? a historic analysis of different types of laughter. In: 10th Congress of the Swiss Society of Psychology. University of Zurich, Switzerland

67. Ivaldi S, Anzalone SM, Rousseau W, Sigaud O, Chetouani M (2014) Robot initiative in a team learning task increases the rhythm of interaction but not the perceived engagement. Front Neurorobot 8:5

68. Izard C (1994) Innate and universal facial expressions: evidence from developmental and cross-cultural research. Psychol Bull 115:288–299

69. Kendon A (1990) Conducting interaction: patterns of behavior in focused encounters (Studies in interactional sociolinguistics). Cambridge University Press, New York

70. Kopp S, Krenn B, Marsella S, Marshall AN, Pelachaud C, Pirker H, Thórisson KR, Vilhjálmsson HH (2006) Towards a common framework for multimodal generation: the behavior markup language. Proceedings of the 6th international conference on Intelligent virtual agents. Springer, Heidelberg, IVA, pp 205–217

71. Kopp S (2010) Social resonance and embodied coordination in face-to-face conversation with artificial interlocutors. Speech Commun 52(6):587–597

72. Kopp S, Gesellensetter L, Krämer NC, Wachsmuth I (2005) A conversational agent as museum guide design and evaluation of a real-world application. In: Panayiotopoulos T, Gratch J, Aylett R, Ballin D, Olivier P, Rist T (eds) Intelligent virtual agents, Lecture notes in computer science, vol 3661, Springer, Berlin, pp 329–343. doi:10.1007/11550617_28

73. Lambertz K (2011) Back-channelling: The use of yeah and mm to portray engaged listenership. Griffith working papers in pragmatics and intercultural communication. vol 4, pp 11–18

74. Langlet C, Clavel C (2014) Modelling user's attitudinal reactions to the agent utterances: focus on the verbal content. In: Proceedings of 5th international workshop on corpora for research on emotion, sentiment & social signals (ES3), Reykjavik, Iceland

75. Langlet C, Clavel C (2015) Improving social relationships in face-to-face human-agent interactions: when the agent wants to know users likes and dislikes. In: Proceedings of annual meeting of the association for computational linguistic

76. Lee J, Marsella S (2006) Nonverbal behavior generator for embodied conversational agents. In: Gratch J, Young M, Aylett R, Ballin D, Olivier P (eds), Proceedings 6th international conference intelligent virtual agents (IVA) Marina Del Rey, CA Springer, LNCS, vol 4133, Aug 21–23 2006, pp 243–255

77. Litman DJ, Forbes-Riley K (2006) Recognizing student emotions and attitudes on the basis of utterances in spoken tutoring dialogues with both human and computer tutors. Speech Commun 48(5):559–590
78. Lombard M, Weinstein L, Ditton T (2011) Measuring telepresence: the validity of the temple presence inventory (tpi) in a gaming context. In: Proceedings of 2011 Annual conference of the International Society for Presence Research (ISPR), ISPR 2011
79. Maat M, Heylen D (2009) Turn management or impression management? In: Proceedings of the 9th international conference on intelligent virtual agents, Springer, Heidelberg, IVA, pp 467–473. doi:10.1007/978-3-642-04380-2_51
80. Maat MT, Truong KP, Heylen D (2010) How turn-taking strategies influence users' impressions of an agent. In: Allbeck J, Badler N, Bickmore T, Pelachaud C, Safonova A (eds) Intelligent virtual agents, Springer, Heidelberg, Lecture notes in computer science, vol 6356, pp 441–453. doi:10.1007/978-3-642-15892-6_48
81. Magnusson M (2000) Discovering hidden time patterns in behavior: T-patterns and their detection. Behav Res Meth, Instrum Comput 32(1):93–110
82. Martin JR, White PR (2005) The language of evaluation. Palgrave Macmillan Basingstoke, New York
83. Martin JR, White PR (2005) The Language of Evaluation. Appraisal in English, Palgrave Macmillan Basingstoke, New York
84. Mehrabian A (1980) Basic dimensions for a general psychological theory: implications for personality, social, environmental, and developmental studies. Oelgeschlager, Gunn & Hain, Cambridge
85. Mol L, Krahmer E, Maes A, Swerts M (2012) Adaptation in gesture: converging hands or converging minds? J Mem Lang 66(1):249–264
86. Morency LP, de Kok I, Gratch J (2010) A probabilistic multimodal approach for predicting listener backchannels. Auto Agents Multi-Agent Syst 20(1):70–84
87. Munezero MD, Suero Montero C, Sutinen E, Pajunen J (2014) Are they different? affect, feeling, emotion, sentiment, and opinion detection in text. IEEE Trans Affect Comput 5:101–111
88. Nakano YI, Ishii R (2010) Estimating user's engagement from eye-gaze behaviors in human-agent conversations. In: Proceedings of the 15th international conference on intelligent user interfaces, ACM, New York, IUI, pp 139–148. doi:10.1145/1719970.1719990
89. Nicolle J, Rapp V, Bailly K, Prevost L, Chetouani M (2012) Robust continuous prediction of human emotions using multiscale dynamic cues. In: Proceedings of the 14th ACM international conference on multimodal interaction, pp 501–508
90. Niederhoffer KG, Pennebaker JW (2002) Linguistic style matching in social interaction. J Lang Soc Psychol 21(4):337–360
91. Niewiadomski R, Pelachaud C (2007) Model of facial expressions management for an embodied conversational agent. In: Proceedings of 2nd international conference on affective computing and intelligent interaction ACII, Lisbon
92. Niewiadomski R, Hyniewska SJ, Pelachaud C (2011) Constraint-based model for synthesis of multimodal sequential expressions of emotions. IEEE Trans Affect Comput 2(3):134–146
93. Niewiadomski R, Demeure V, Pelachaud C (2010) Warmth, competence, believability and virtual agents. In: Allbeck J, Badler N, Bickmore T, Pelachaud C, Safonova A (eds) Intelligent virtual agents, Lecture notes in computer science, vol 6356, Springer, Heidelberg, pp 272–285. doi:10.1007/978-3-642-15892-6_29
94. Niewiadomski R, Obaid M, Bevacqua E, Looser J, Anh LQ, Pelachaud C (2011b) Cross-media agent platform. In: Proceedings of the 16th international conference on 3D web technology, ACM, New York, Web 3D, pp 11–19. doi:10.1145/2010425.2010428
95. Ochs M, Prepin K, Pelachaud C (2013) From emotions to interpersonal stances: Multi-level analysis of smiling virtual characters. In: Proceedings of 2013 humaine association conference on affective computing and intelligent interaction (ACII), IEEE, pp 258–263
96. Osherenko A, Andre E, Vogt T (2009) Affect sensing in speech: Studying fusion of linguistic and acoustic features. In: Proceedings of affective computing and intelligent interaction (ACII)

97. Paleari M, Lisetti C (2006) Psychologically grounded avatars expressions. In: Proceedings of first workshop on emotion and computing at KI, 29th annual conference on artificial intelligence, Bremen, Germany
98. Pardo JS (2006) On phonetic convergence during conversational interaction. J Acoust Soc Am 119(4):2382–2393
99. Pecune F, Cafaro A, Chollet M, Philippe P, Pelachaud C (2014) Suggestions for extending saiba with the vib platform. In: Workshop on architectures and standards for IVAs, held at the '14th international conference on intelligent virtual agents (IVA)', Bielefeld eCollections, pp 16–20
100. Pecune F, Mancini M, Biancardi B, Varni G, Volpe G, Ding Y, Pelachaud C, Camurri A (2015) Laughing with a virtual agent. In: Proceedings of the 14th international joint conference on Autonomous agents and multiagent systems
101. Pedica C, Vilhjálmsson HH (2010) Spontaneous avatar behavior for human territoriality. Appl Artif Intell 24(6):575–593. doi:10.1080/08839514.2010.492165
102. Pekrun R, Cusack A, Murayama K, Elliot AJ, Thomas K (2014) The power of anticipated feedback: effects on students' achievement goals and achievement emotions. Learn Instruct 29:115–124
103. Pelachaud C, Badler N, Steedman M (1996) Generating facial expressions for speech. Cognit Sci 20(1):1–46
104. Perrin L, Deshaies D, Paradis C (2003) Pragmatic functions of local diaphonic repetitions in conversation. J Pragmat 35(12):1843–1860
105. Peters C, Castellano G, de Freitas S (2009) An exploration of user engagement in hci. In: Proceedings of the international workshop on affective-aware virtual agents and social robots, ACM, p 9
106. Pickering MJ, Garrod S (2004) Toward a mechanistic psychology of dialogue. Behav brain sci 27(02):169–190
107. Plutchnik R (1980) Emotion: a psychoevolutionary synthesis. Harper and Row, NY
108. Poggi I (2007) Mind, hands, face and body: a goal and belief view of multimodal communication. Weidler, Berlin
109. Prepin K, Ochs M, Pelachaud C (2013) Beyond backchannels: co-construction of dyadic stance by reciprocal reinforcement of smiles between virtual agents. In: International conference CogSci (Annual conference of the cognitive science society)
110. Provine RR (2001) Laughter: a scientific investigation. Penguin, New York
111. Ravenet B, Ochs M, Pelachaud C (2013) From a user-created corpus of virtual agents non-verbal behavior to a computational model of interpersonal attitudes. In: Aylett R, Krenn B, Pelachaud C, Shimodaira H (eds) Intelligent virtual agents, Lecture notes in computer science. Springer, Heidelberg, pp 263–274. doi:10.1007/978-3-642-40415-3_23
112. Ravenet B, Ochs M, Pelachaud C (2013) From a user-created corpus of virtual agent's non-verbal behavior to a computational model of interpersonal attitudes. In: Proceedings 13th international conference intelligent virtual agents, IVA 2013, Edinburgh, UK, August 29–31 2013, pp 263–274
113. Ravenet B, Cafaro A, Ochs M, Pelachaud C (2014) Interpersonal attitude of a speaking agent in simulated group conversations. In: Bickmore T, Marsella S, Sidner C (eds) Intelligent virtual agents, Lecture notes in computer science, vol 8637, Springer International Publishing, pp 345–349. doi:10.1007/978-3-319-09767-1_45
114. Ravenet B, Cafaro A, Ochs M, Pelachaud C (2014) Interpersonal attitude of a speaking agent in simulated group conversations. In: Proceedings of 14th international conference on intelligent virtual agents, IVA 2014, Boston, August 27–29, 2014, pp 345–349
115. Reynolds C (1999) Steering behaviors for autonomous characters. Miller Freeman Game Groups, San Francisco, Proceedings of the game developers conference, p 763
116. Rich C, Sidner CL (2012) Using collaborative discourse theory to partially automate dialogue tree authoring. In: Proceedings of the 12th international conference on intelligent virtual agents, Springer, Heidelberg, IVA'12, pp 327–340. doi:10.1007/978-3-642-33197-8_34
117. Russell J (1980) A circumplex model of affect. J Person Soc Psychol 39:1161–1178

118. Ruttkay Z, Noot H, ten Hagen P (2003) Emotion disc and emotion squares: tools to explore the facial expression face. Comput Graph Forum 22(1):49–53
119. Scherer K (2000) Emotion. In: Stroebe W, Hewstone M (eds) Introduction to social psychology, a European perspective. Oxford Blackwell Publishers, Oxford, pp 151–191
120. Scherer K (2005) What are emotions? And how can they be measured? Soc sci inf 44(4):695–729
121. Schmitt A, Polzehl T, Minker W (2010) Facing reality: simulating deployment of anger recognition in IVR systems. In: Lee GG, Mariani J, Minker W, Nakamura S (eds) Spoken dialogue systems for ambient environments. Lecture notes in computer scienceSpringer, Heidelberg, pp 122–131
122. Schroder M, Bevacqua E, Cowie R, Eyben F, Gunes H, Heylen D, Ter Maat M, McKeown G, Pammi S, Pantic M et al (2012) Building autonomous sensitive artificial listeners. IEEE Trans Affect Comput 3(2):165–183
123. Schuller B, Rigoll G, Lang M (2004) Speech emotion recognition combining acoustic features and linguistic information in a hybrid support vector machine-belief network architecture. Proceedings of ICASSP, Montreal 1:577–580
124. Schuller B, Batliner A, Steidl S, Seppi D (2011) Recognising realistic emotions and affect in speech: state of the art and lessons learnt from the first challenge. Speech Commun 53(9):1062–1087
125. Sidner CL, Dzikovska M (2002) Human-robot interaction: engagement between humans and robots for hosting activities. In: Proceedings of the 4th IEEE international conference on multimodal interfaces, IEEE Computer Society, p 123
126. Sidner CL, Lee C, Kidd CD, Lesh N, Rich C (2005) Explorations in engagement for humans and robots. Artif Intell 166(1):140–164
127. Silverman LH (1999) Meaning making matters: communication, consequences, and exhibit design. Exhibitionist
128. Smith C, Crook N, Dobnik S, Charlton D (2011) Interaction strategies for an affective conversational agent. Presence: teleoperators and virtual environments. MIT Press, Cambridge, pp 395–411
129. Srikant R, Agrawal R (1996) Mining sequential patterns: generalizations and performance improvements. Adv Database Technol 1057:1–17
130. Stivers T (2008) Stance, alignment, and affiliation during storytelling: when nodding is a token of affiliation. Res Lang soc interact 41(1):31–57
131. Svennevig J (2004) Other-repetition as display of hearing, understanding and emotional stance. Discourse studies 6(4):489–516
132. Takashima K, Omori Y, Yoshimoto Y, Itoh Y, Kitamura Y, Kishino F (2008) Effects of avatar's blinking animation on person impressions. In: Proceedings of graphics interface, Toronto, Ontario, Canada, GI '08, Canadian Information Processing Society, pp 169–176. doi:10.1145/1375714.1375744
133. Tannen D (1992) Talking voices: repetition, dialogue, and imagery in conversational discourse. cambridge University Press, Cambridge
134. Traum D, Aggarwal P, Artstein R, Foutz S, Gerten J, Katsamanis A, Leuski A, Noren D, Swartout W (2012) Ada and grace: Direct interaction with museum visitors. In: Nakano Y, Neff M, Paiva A, Walker M (eds) Intelligent virtual agents, Lecture notes in computer science, vol 7502, Berlin Heidelberg, Springer, pp 245–251. doi:10.1007/978-3-642-33197-8_25
135. Truong KP, Poppe R, Heylen D (2010) A rule-based backchannel prediction model using pitch and pause information, pp 3058–3061
136. Tsapatsoulis N, Raouzaiou A, Kollias S, Cowie R, Douglas-Cowie E (2002) Emotion recognition and synthesis based on MPEG-4 FAPs in MPEG-4 facial animation. In: Pandzic IS, Forcheimer R (eds) MPEG4 facial animation–the standard, implementations and applications. Wiley, New York
137. Vilhjálmsson HH, Cantelmo N, Cassell J, E Chafai N, Kipp M, Kopp S, Mancini M, Marsella S, Marshall AN, Pelachaud C, Ruttkay Z, Thórisson KR, Welbergen H, Werf RJ (2007) The behavior markup language: recent developments and challenges. In: Proceedings of the

7th international conference on intelligent virtual agents, Heidelberg, IVA '07, Springer, pp 99–111
138. Vinciarelli A, Pantic M, Bourlard H (2009) Social signal processing: survey of an emerging domain. Image Vis Comput 27(12):1743–1759
139. Wang N, Johnson WL, Mayer RE, Rizzo P, Shaw E, Collins H (2008) The politeness effect: pedagogical agents and learning outcomes. Int J Hum-Comput Studies 66(2):98–112
140. Ward A, Litman D (2007) Automatically measuring lexical and acoustic/prosodic convergence in tutorial dialog corpora. In: Proceedings of the SLaTE Workshop on speech and language technology in education
141. With S, Kaiser S (2011) Sequential patterning of facial actions in the production and perception of emotional expressions. Swiss Journal of Psychology / Schweizerische Zeitschrift fr Psychologie / Revue Suisse de Psychologie 70(4):241–252
142. Yildirim S, Narayanan S, Potamianos A (2011) Detecting emotional state of a child in a conversational computer game. Comput Speech Lang 25(1):29–44
143. Zhang L (2009) Exploration of affect sensing from speech and metaphorical text. In: learning by playing. Game-based education system design and development, springer, Berlin, pp 251–262
144. Zhao R, Papangelis A, Cassell J (2014) Towards a dyadic computational model of rapport management for human-virtual agent interaction. Intelligent virtual agents. Springer, Berlin, pp 514–527

Chapter 8
Virtual Coaches for Healthy Lifestyle

H.J.A. op den Akker, R. Klaassen and A. Nijholt

Abstract Since the introduction of the idea of the software interface agent the question recurs whether these agents should be personified and graphically visualized in the interface. In this chapter we look at the use of virtual humans in the interface of healthy lifestyle coaching systems. Based on theory of persuasive communication we analyse the impact that the use of graphical interface agents may have on user experience and on the efficacy of this type of persuasive systems. We argue that research on the impact of a virtual human interface on the efficacy of these systems requires longitudinal field studies in addition to the controlled short-term user evaluations in the field of human computer interaction (HCI). We introduce Kristina, a mobile personal coaching system that monitors its user's physical activity and that presents feedback messages to the user. We present results of field trials ($N = 60$, 7 weeks) in which we compare two interface conditions on a smartphone. In one condition feedback messages are presented by a virtual animated human, in the other condition they are displayed on the screen in text. Results of the field trials show that user motivation, use context and the type of device on which the feedback message is received influence the perception of the presentation format of feedback messages and the effect on compliance to the coaching regime.

H.J.A. op den Akker · R. Klaassen · A. Nijholt (✉)
Human Media Interaction, University of Twente,
217, 7500 AE Enschede, The Netherlands
e-mail: a.nijholt@utwente.nl

H.J.A. op den Akker
email: h.j.a.opdenakker@utwente.nl

R. Klaassen
e-mail: r.klaassen@utwente.nl

© Springer International Publishing Switzerland 2016
A. Esposito and L.C. Jain (eds.), *Toward Robotic Socially Believable
Behaving Systems - Volume II*, Intelligent Systems Reference Library 106,
DOI 10.1007/978-3-319-31053-4_8

8.1 Introduction

We are at the coffee corner of Jones' office. Jones has a tough day and he is in need now of another coffee. Facing the machine Jones routinely presses the GO button. The machine says "select your item" and Jones presses the coffee-with-sugar button. When the machine requests authorization he holds his card in front of the machine's card reader. Jones expects that the machine will now ask him to wait a while. Instead, it says: "Mr. Jones, may I remind you that this is your fourth coffee today. The weather is nice. A short walk will do you well. That is healthier than another coffee."

Clearly, Jones' coffee machine is not just a coffee machine. It is integrated into a pervasive system that also keeps record of his daily consumptions. It uses information about his physical activity, that it has read from his step counter, a built-in function of his smart watch. It knows about the local weather. The system uses all this information to carry out its task which is not just to serve coffee on demand but to care about mr. Jones' well being. The machine has become a representative agent of a system that tries to nudge its clients to live a more healthy life. This is an example of a *behavior change support system* or personal coaching system. It uses persuasive strategies and principles from captology, the study of persuasive technology, to bring about its intended effect on its users.

In their 1993 paper "Learning Interface Agents" Pattie Maes and Robyn Kozierok describe an *intelligent interface agent* that learns how to sort messages into the different mailboxes created by the user. The authors use the metaphor of the "personal assistant" that is collaborating with the user in the same work environment [41]. The interface agent learns by continuously "looking over the shoulder" when the user is performing actions (p. 461). It stores situations (information about incoming e-mail) with the actions performed by the user in these situations. When a new mail arrives the agent suggests an action based on the similarity between the new situation and previously seen cases. The agent also learns from the user who gives either explicit feedback, e.g., by correcting the agent's action, or implicit feedback, e.g., by ignoring a suggestion by the agent.

Since the seventies of the 20th century we witness several changes in the paradigm for human computer interaction and in the metaphors we use when we think and talk about interacting with computers. First, the computer changed from a computing device that has to be programmed by expert users and scientists through a command line interface to a communication medium that could also be used by nonexpert users. Alan Kay's Dynabook, designed in 1968, was intended for children! [35]. From the desktop metaphor according to which the user is "navigating the information space" by means of graphical items, icons representing administrative functions of the computer, the paradigm shifted to that of the conversation. Today it is quite common to think about the computer as a social agent. The personal computer has become a person itself.

The discussion between proponents of interface design methods within the intelligent agent paradigm, e.g., Pattie Maes, and opponents of this paradigm, e.g., Ben Schneiderman, advocates of the direct manipulation or desktop metaphor, boils down to a common concern about who is in control of the system and who is responsible for what the system does [61]. Systems become more and more complex and proponents argue that interface agents that understand the user and act in a proactive way make systems accessible to nonexpert users. The interface should provide intuitive means of natural interaction and software agents should act on behalf of their users. Opponents see the danger of delegating tasks to proactive autonomous software agents. The system may run out of control of the user. The problem is who is responsible then.[1]

A difference between the personal assistant of Kozierok and Maes [41] and the social caring agent behind Jones' coffee machine is in the types of work spaces in which the collaboration between system and user takes place. Maes' personal assistant is concerned with a very specific information processing task. The system behind Jones' coffee machine tries to support Jones to change his sedentary life style -typical for most office workers that look at computer screens for many hours a day- into a more healthy one. Daily personal life has become a design space. Machine learning techniques such as the one Maes and Kozierok implemented in their personal assistant are also used in Jones' lifestyle change support system. After some time it will learn if Jones likes to be mothered by the system. The acceptance of this type of personal persuasive systems is a real issue. The feeling of being in control appears to be an important factor.

Using interface agents does not necessarily mean that they are anthropomorphized and that they should be graphically presented as a human in the interface. It is, however, interesting to see that the graphical interface of the interactive learning assistant contains a caricature that conveys the functional and cognitive state of the agent to the user. Where computers make computations of which only the outcome is interesting for the user, the personal assistant approaches the user in a social way: it asks questions, suggests solutions, it even tries to persuade the user to take a break instead of having another coffee. As soon as the technology is there to graphically represent the agent on the interface it seems hard to resist to do so.

---

[1]From the very beginning of the rise of captology ethical implications of persuasive technology have been considered. Fogg [26] devotes a chapter to it. Heckman and Wobbrock [31] point at the fact that "anthropomorphic agents provide electronic commerce consumers with terrific benefits yet also expose them to serious dangers". They conclude that developers must focus agent design on consumer welfare, not technical virtuosity, if legal and ethical perils are to be avoided. In a special issue of the Communications of the ACM [7] present ethical principles of persuasive system design in a framework in which they model persuasive technology as an instrument between the persuader and the persuadee. Based on new insights in the phenomenology of technology [66] considers persuasive technology as an example of technological mediation. New technologies reshape human practices and views on reality. People don't need cars or internet when cars or internet have not been invented. Ethical concerns are not restricted to "intended persuasions" and "unintended outcomes" of the use of technology. For example, persuasive technology made it possible to frame the act of Jones having a coffee within a framework of healthy living and makes him aware of the fact that the system is looking over his shoulder; maybe more as a big brother than as a caring sister. Being a stakeholder in technology development researchers and designers of new technologies have special responsibilities for the mediating implications of these technologies.

Advances in computer graphics made it possible to build *animated graphical interface agents*. Such agents appear as embodied characters that exhibit lifelike human behaviors and that convey affective state and communicative intentions to the user by gesturing and gaze behavior [24]. The animated embodiment is often just the visual appearance of an *embodied conversational agent* (ECA) [21]. Under the hood the ECA is a more or less sophisticated (spoken or multi-modal) dialogue system. Cassell [19] advocates the use of ECA interfaces and to represent a system as a human in those cases where social collaborative behavior is key. The purpose of the ECA is "to leverage users' knowledge of how to conduct face-to-face conversation" and to "leverage users' natural tendencies to attribute humanness to the interface." (p. 82).

The terms "virtual human", "embodied conversational agent" or "intelligent virtual agent" for "computer models resembling humans in their bodily look and their communication capabilities" [59] are often used interchangeably. Virtual humans can however play quite different roles in research and in systems. Depending on the interest of the researcher virtual humans are built and studied as simulation of real humans with a certain personality (e.g., [13]) or as animated interface agents. The term *interface* agent suggests that the function of the virtual human is that of communication channel between a user and the system. When a virtual human plays—for example—the role of a suspect in a serious game for interrogative interview training the social interaction itself with the simulated suspect is what the system is made for. Hence, it depends on the specific function that the virtual human has what graphical, behavioral, cognitive and affective features need to be implemented.

Technology aided coaching on healthy behavior is widely regarded as a promising paradigm to aid in the prevention of chronic diseases, in the adherence to medical treatment and in the process of healthy ageing in general [33]. In order to encourage physical activity in patients suffering from chronic diseases, as well as healthy adults, many different coaching systems have been developed. Typically these consist of an activity sensor and some form of coaching application delivered either through a web portal, smartphone or through the sensor itself [53]. These personal coaching systems more and more try to make use of the progress made in the field of natural interfaces, spoken dialogue systems and embodied conversational agents. Since the human coach is paradigmatic for the personal digital coach it lies at hand to present the digital coach by an (animated) virtual human, a talking head or an ECA [36].

In this chapter we look at the use of these embodied conversational agents as a natural interface of persuasive systems that care for their user's well being and health. Does the use of an ECA add to the acceptance, the usability, the efficacy, the persuasiveness, the attractiveness of these type of systems? And if they do, is it just because of the (animated) graphical appearance of a face or is it because of the naturalness of the interaction made possible by the speech interface? Or, are other factors into play that influence these outcome measures?

An important issue is the method for evaluating the effects of the interface design on the efficacy of behavior change support systems. Experience with therapies for patients with chronic conditions (low back pain, COPD, diabetes) show that adherence to the therapy in the first weeks is no guarantee for adherence on the longer term. Thus to measure the efficacy of coaching systems long-term experiments seem to be required. In research environments most user evaluation studies are performed with early stage technologies and through short-term lab experiments. Research has shown that the use of animated graphical agents often raises the entertainment value of a system. Dehn and van Mulken [24] point at the fact that short-term user studies might not be appropriate to measure the long-term effect on utilitarian performance measures. Although the entertainment value of the interface design may motivate the user to seek interaction with the system, in the long run usefulness is fundamental for long-term user engagement.

Klasnja et al. [39] argue that "behavior change in the traditional clinical sense is not the right metric for evaluating early stage technologies that are developed in the context of HCI research." They argue that because of the complexity of behavior change support systems randomized control trials are not suitable to answer the question *what* exactly causes the outcome of such trials. Instead HCI evaluation methods should be developed that focus on the efficacy of *specific persuasive strategies* implemented in the persuasive system, such as for example self-monitoring, or social dialog. The question we focus on here is how the use of a virtual human in the interface impacts the effect of certain persuasive strategies used.

To find an answer to this question we will first in Sect. 8.2 review what type of persuasive features and strategies are included in persuasive systems. In Sect. 8.3 we will review research findings concerning the effects of virtual humans as interface agents of systems in which perceived credibility and persuasiveness play a role. In Sect. 8.4 we will present Kristina, a digital coaching system that supports and motivates people to live a balanced and healthy lifestyle by providing feedback about their level of physical activity [52]. Kristina is designed as a multi-device (PC, Mobile, TV) system that is able to monitor physical activity and medication intake of its users.[2] It implements a number of persuasive strategies and the virtual human used has some features to improve persuasiveness. In Sect. 8.5 we present a six week user evaluation study of the Kristina coaching system in which we compare two different ways of presenting coaching messages, with or without the use of an ECA in the interface. In Sect. 8.6 we present lessons learned regarding the use of virtual humans as user interface in this type of persuasive systems.

8.2 Persuasive Communication and Persuasive Technology

Will Jones follow the advice of the coffee machine and take a break instead of having another coffee? A change in someone's opinion, attitude or behavior is often the result of persuasion, in which someone, who has the role of *persuader*, tries to influence the *persuadee* through a process of communication. In this section we

---

[2]The system is built in the European ARTEMIS project on Smart Composite Human-Computer Interfaces (Smarcos) that focused on inter-usability aspects of multi-devise/multi-sensor service systems (2010–2012). https://artemis-ia.eu/project/24-smarcos.html.

review the most important theories of persuasive communication and persuasive technology. We do that in order to see in the next section in what way the use of virtual humans in the interface of a persuasive system could impact the persuasiveness of the communication.

8.2.1 Persuasive Communication

The process of persuasive communication has four elements each with a number of aspects [9]:

1. Message: timing, the argument, the emotional value
2. Source: expertise, trustworthiness, credibility, attractiveness (liking, similarity), role model, power
3. Channel: media types (newspaper, tv, internet), devices, modalities (visualisation, spoken language)
4. Recipient: intelligence, self-esteem, state of change, motivation, needs, personality

Petty and Cacioppo [56] developed a model of how recipients elaborate on the message they receive. According to this Elaboration Likelihood Model there are two routes for persuasion to behavior and attitude change: the central route, where the receiver elaborates on the (argumentative or affective) content of the message, and the peripheral route where the receiver responses to peripheral cues. The way the receiver elaborates on the message and involves himself in the content can vary. Internal processes on the side of the recipient influence the persuasiveness of the communication (see Fig. 8.1): attention, comprehension, yielding, and retention.

Regarding the message content: does it matter for the persuasiveness of the argument if the message brought by the persuader contains evidence -statements referring to others than the persuader to support his claim? In a study with students as audience of two speakers on two different topics [47] found that although good use of evidence increases perceived authoritativeness, it does not appear to have an effect on perceived character. Topic and persuader's gender may however mediate the effect of evidence use on the perceived character. Moreover, speeches which include good use of evidence produce a greater attitude shift than those which do not.

But often receivers of a persuasive message are not motivated to elaborate on the content of the message. Instead, they evaluate and respond to it by looking at some peripheral cues. Common used cues are related to *source credibility* (we follow a suggestion because we believe it comes from a credible source), to *liking* (the persuader is an attractive person) and to *social consensus* (we believe, do and need what our social peers believe, do and need).

In order to be successful in bringing about a change in the persuadee's behavior receiver and persuasive communication must satisfy some requirements. The Transtheoretical Model of Behavior [57] is a popular model that describes the process people go through when changing their behavior. According to Prochaska change in

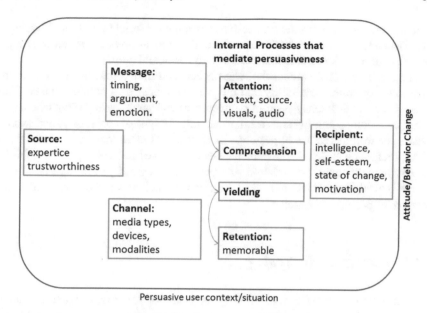

Fig. 8.1 A model of persuasive communication (based on [55])

behavior can be seen as a process involving different stages: (1) pre-contemplation, (2) contemplation, (3) preparation (or determination), (4) action and (5) maintenance. A persuader, no matter whether it is a human or a persuasive system, should adept it's intervention strategy to the persuadee's state of change. When Jones is already motivated to be more physical active a motivational interview may not be the effective way to intervene. The self-efficacy theory by Bandura [2] is integrated into the Transtheoretical Model. Self-efficacy is the confidence people have in their own ability to deal with specific situations without returning to old negative behaviors. Self-efficacy is strengthened naturally through success. A behavior change support system can make users explicitly aware of their successes by showing them monitored data and by giving them positive feedback.

The persuasive behavior model of Fogg [27] provides three factors that determine if people can be persuaded to change their behavior: (1) Motivation, (2) Abilities, and (3) Triggers. The model asserts that for a target behavior to happen, a person must have sufficient motivation, sufficient ability to perform the intended behavior, and an effective trigger. All these three factors should be present for the targeted behavior to occur. The motivation of a user and ability to perform the targeted behavior define the chance that the user will actually perform the targeted behavior. The timing of the triggers is an important issue. According to the principle of Kairos a trigger such as a suggestion to do some action should be presented at the most opportune moment. Jones' coffee machine works according this principle when it suggests the user to take a break instead of having another coffee at the moment he is ordering a coffee. The principle of Kairos is one of the principles for message tailoring [54]. Fogg's three factors also occur in the Elaboration Likelihood Model in Petty and Cacioppo [56]. Dependent on the level of attention and the motivation of the receiver a trigger can work through the central route or through peripheral cues.

Burgoon et al. [16] studied the relation between non-verbal behaviors, several dimensions of source credibility and the persuasiveness of a public speaker. They found that speaker fluency (sign of competence) and pitch variety (sign of engagement), eye contact, smiling and facial pleasantness (signs of sociability) and expressiveness as well as the use of gestures, in particular illustrators (i.e. gestures timed with speech, such as deictic movements, head nods, eye blinks as sign of sociability) had a significant positive correlation with perceived credibility of the speaker. These are the typical "behaviors" that virtual humans exhibit when they are used in the interface of persuasive systems.

8.2.2 Persuasive Technology

Fogg [26] defined persuasive technology as technology that is designed to change attitudes or behaviors of the users through persuasion and social influence, but not through coercion. According to Fogg there are different strategies that can be applied in persuasive technology:

- Reduction—Using computer technology to reduce complex behavior into simple tasks. Reduction makes it easier to perform a behavior and will increase the benefit/cost ratio of the behavior. A better benefit/cost ratio increases the motivation of a user to engage in the behavior.
- Tunneling—Using computer technology to guide users through the process of behavior change. During this process the technology will create opportunities to persuade their user.
- Tailoring—Using computer technology to present (persuasive) information tailored to the needs, interests, personality and context of the user.
- Suggestion—Using computer technology to present suggestions to the user to act in a certain way. Suggestions presented at the right moment in time will be more effective.
- Self-monitoring—Using computer technology to automate the tracking of behavior of the user. Self-monitoring helps people to achieve predetermined goals or outcomes.
- Surveillance—Using computer technology to monitor the behavior of other users in order to change your own behavior.
- Conditioning—Using computer technology for positive reinforcement to change existing behavior into habits or to shape complex behaviors.

Although Fogg's model of behavior does suggest design principles for persuasive technology, it does not explain how these principles should be transformed into software requirements and system features. Based on the principles of persuasive communication [51] provided a Persuasive System Design (PSD) framework for designing and evaluating persuasive systems. PSD presents four categories of system features; (1) primary task support, (2) dialogue, (3) system credibility and (4) social support.

Primary task support features correspond with the above mentioned persuasive strategies of Fogg. The design principles related to the dialogue category are about the user interaction with the system. Computer-human dialogue includes praise, rewards, reminders and suggestions. The third category of system features is related to system credibility. The system must provide information that is truthful, fair and unbiased and it must incorporate **expertise** by showing knowledge, experience and competence to increase persuasiveness. The fourth and last category is related to social support. Design principles in this category are based on the idea of designing a system that motivates users by leveraging social influence. Design principles in this category include **social learning**, people will be more motivated to perform a targeted behavior when he or she can observe the behavior of others who are using the system or when they can compare their own results with other users.

8.2.3 The Computer as Persuasive Social Actor

The metaphor of the computer as social actor is supported by experimental research that shows that users treat the computer they work with in a way they treat other human subjects [58]. This works on a subconscious level. If we ask people they will say they do not consider the computer as a social agent: the computer is just a machine. But instinctively people behave as if they are interacting with a human. Social cues can be divided into five primary types:

- Physical—such as having a face, body, eyes or movement
- Psychological—such as showing personality, feelings or empathy
- Language—such as interactive language use or spoken language
- Social dynamics—such as turn taking, cooperation, praise for good work or answering questions
- Social roles—such as doctor, coach, team mate, opponent, teacher or pet

Embodied conversational agents offer many opportunities to support the presentation of a persuasive system as a social actor. The working of the interface design is largely based on social psychology. By simulating natural signs that we know from human-human social interaction the interface designer tries to elicit user experiences that make him behave in a similar way he does in the real case. HCI designers do not do that because they want to deceive the user (they had to deceive themselves then) or for artistic reasons (although aesthetics matters), but primarily because simulation works; it is effective. A very commonly known user experience occurs when we read or hear a text. When Jones' coffee machine "says" "Please show your card" Jones immediately experiences this as a request because it reminds him of similar signals in similar situations. In the following section we will review experimental research that aims to identify those socio-psychological mechanisms in the use of virtual humans that impact the various factors that influence the efficacy of persuasive communication by means of persuasive technology.

8.3   The Persuasiveness of Virtual Humans

Since the introduction of the idea of using virtual humans or embodied conversational agents (ECA) as interface agent in the field of HCI, the question recurs "whether software agents should be personified in the interface" [20, 40]. In 1996 Koda and Maes state that "the goal of HCI work should be to understand when a personified interface is appropriate". Many researchers examined the presence and the type of embodied agents but there is little consensus as to whether or not the presence of visual agents improves a users experience with an interface, and if so by what degree [70].

Does the use of a virtual human in the interface affect the persuasiveness of personal coaching systems? Or, rather: what kinds of animated interface agent and which kind of behavioral and expressive features in what task domain and use context have a positive impact on the persuasiveness of the system? The first question is the question of *presence*: in what cases is it a good idea to have a graphical representation of the interface agent and if so how should it be presented; e.g. how realistic: a 2D or 3D caricature, photographic, by means of a talking head or full-body, animated, or as a physical robot? Then the question is how it should look like and what types of behaviors impact what factors that influence persuasiveness.

We have seen that source credibility (believability, trustworthiness) is an important factor for persuasiveness. Research in human communication has shown that eye contact rates influences credibility. To be persuasive a speaker should engage in more eye contact. But does this hold for ECAs as well? There are several reasons why we must be cautious before we jump to this conclusion. The communication channel has effect on the communication. It is not necessarily so that gaze of an ECA in the direction of the user is perceived similar as eye contact is perceived by a human persuader in a face-to-face encounter. Moreover, many variables describing behavior of the persuader jointly affect credibility and -indirectly- persuasiveness. Many studies in social psychology and HCI try to isolate the influence of one or a pair of variables in a controlled experimental setting. But this does not mean that the same effect happens in a different setting. The relation between the complex structure of variables on the one hand and the perceived credibility of the source on the other hand is non-linear and unstable so that a small change in the situation may make the difference in how the recipient perceives the source's credibility. (Compare the relation between a scribble and the word or picture that the reader recognizes in it: a small change may result in a completely different perceived word or picture.) That being said we will present in this section results from studies that provide at least indications of factors that influence the persuasiveness of a virtual human in the interface.

8.3.1 Naturalness and Credibility

Yee et al. [70] examined a number of theoretical questions related to interfaces: the effect of the presence of an embodied agent, how **realistic** the agent appears through animations and behaviors, and the type of response which was measured (subjective or performance). Do people react differently to interfaces with (1) no visual representation, (2) a human-like representation with low realism (e.g., cartoon figure), and (3) a human-like representation with high realism (e.g., 3D model animated with gestures)? Realism is not the same as anthropomorphism, which is essentially the perception of humanlike characteristics in either real or imagined nonhuman agents or objects [25]. There are many dimensions on which an interface agent can be considered real. It can behave realistically through animations, it can be highly photographically realistic via computer graphics, or alternatively it can be highly human like (i.e., anthropomorphic) [13]. Results by Guadagno et al. [29] supported the model of social influence within immersive virtual environments by Blascovich et al. [13]. Specifically, the prediction that virtual humans high in **behavioral realism** would be more influential than those low in behavioral realism was supported, but this effect was moderated by the gender of the virtual human and the research participant.

Baylor [5] points at the impact that anthropomorphic interface agents have on the motivation of learners. "Together with motivational messages and dialogue, the agents appearance is the most important design feature as it dictates the learners perception of the agent as a virtual social model, in the Bandurian sense. The message delivery, through a human-like voice with appropriate and relevant emotional expressions, is also a key motivational design feature."

Effects of animated agents in the user interface with the intention to simulate a natural and believable way of interaction have been evaluated in many studies [24, 43, 49, 50, 62]. In 1997 Lester et al. [43] posed the *persona effect*. The idea behind the persona effect is that a lifelike agent in a learning environment has a strong positive effect on the user's perception of their learning experience, and conclusions from this research showed improved learning performances. These results were questioned in further research [24]. Adding a virtual human to the user interface can be a source of distraction to the user and may disturb the human computer interaction [8]. Mazzotta et al. [45] showed that feedback messages in the domain of healthy eating were better evaluated when they were presented by a virtual human compared to feedback messages presented in text. Text messages were easier to understand, but messages presented by the virtual human were perceived as more persuasive and reliable. Mazotta concludes by hypothesing that text messages are better suited for simple information given tasks, while more persuasive messages (reflecting the social and emotional intelligence of the virtual human) could be presented by a virtual human to increase the effectiveness of the persuasive strategies. Several studies by Murano et al. [50] showed seemingly inconsistent results in relation to using anthropomorphic user interfaces to present feedback in computer systems. The study shows that acceptance and effectiveness of anthropomorphic interfaces depends on the specific task. Catrambone et al. [22] also suggested that the types of task influences the results of these type of experiments. In the long-term users may assign affective states to animated interface agents that are seen as companions [10]. Turunen et al. [63] developed a multi-modal companion system that built relationship with the user of the system to support everyday health and fitness related activities. User interface agents and support for conversational dialogues should help to build social relationships between the user and the system, and these relationships should motivate towards a healthier lifestyle.

Using a meta-analysis of research in 25 papers reporting empirical studies on the effect of presentation and realism on task performance [70] produced some consistent findings. First, the presence of a representation produced more positive social interactions than not having a representation. This effect was found in studies that used both subjective and behavioral measures. Subjective measures are taken by surveys, interviews and questionnaires. Secondly, **human-like** representations with **higher realism** produced more positive social interactions than representations with lower realism; however, this effect was only found when subjective measures were used. Behavioral measures did not reveal a significant difference between representations of low and high realism. A recent survey of 33 studies by Li [44] compares persuasiveness of physical robots with robots on a screen and with virtual agents. Physical robots score higher in persuasiveness than robots and virtual characters on the screen. There was no difference in user response between the robot on the screen and the virtual character on the screen. This indicates that physical presence more than appearance of the virtual agent affects persuasiveness. Long-term field studies in social robotics are needed to see if these effects persist over time.

Perceived expressions of emotional involvement: signs of caring, empathy and understanding has impact on sociability and thus on credibility. Does synthesized expression of emotion either verbally or non-verbally impact persuasiveness? In a literature review [6] focuses on the question "which kind of emotional expressions in which way influences which elements of the user's perceptions and behaviours?" (p. 757). Berry et al. [8] empirically evaluated an ECA named Greta to investigate the effects of emotion expression in a system that promotes healthy eating habits. The messages from the system could be presented by Greta, by a human actor, only by the voice of Greta or via a text message. Except for the human actor, presenting the message by an ECA received the highest ratings for helpfulness and likability. They found that persuasive health messages appeared to be more effective (measured by subjective measures) when presented with an emotionally neutral expression compared to when messages were presented with an emotion consistent with the content of the message.

8.3.2 *Virtual Humans and Behavior Change Support Systems*

Results from studies by Henkemans et al. [32], Schulman and Bickmore [60], Berry et al. [8] indicate that the use of virtual characters can have a positive effect on the likability, helpfulness, ease of use and motivation to use computer systems.

Bickmore conducted several experiments with the MIT FitTrack system [10, 11, 60]. By using the FitTrack system users can enter their daily steps counted by a pedometer and estimated time of physical activity. Schulman and Bickmore [60] studied the effects of different versions of computer agents on users' attitudes towards regular exercises. This experiment compared two situations, an ECA versus a text agent and whether the agent attempted to build a user-agent relationship through social dialogue in a wizard-of-oz set-up. Participants following the persuasive dialogue showed a significant increase in positive attitude towards the exercises. This change was significantly smaller when the agent used social dialogue. Users' perceptions of the dialogue were most positive for an ECA with social dialogue or a text-only agent without social dialogue.

In a long-term (2 and 12 months) study Bickmore et al. [12] tested the efficacy of an animated virtual exercise coach designed for sedentary older adults with low health literacy. Positive effects of the ECA enhanced system compared to a system that uses only a pedometer measured after 2 month waned after 12 month. This might be caused by the fact that in the ten months period after the first two months interaction was not through a tablet anymore but through a screen in a waiting room kiosk of a clinic.

The MOPET system of Buttussi and Chittaro [17] is a *mobile* personal trainer system. MOPET is designed to support the user throughout exercise sessions, by guiding the user through fitness trails that alternate running with physical exercises. It tracks the user's position on the trail and shows the user's speed, and also tries to motivate the user through messages. It uses an external sensor device that collects heart rate and accelerometer data. When the user comes to an exercise point along the route, the system recognizes this and demonstrates the exercise to the user. The virtual human is presented as a full-bodied animated 3D character that is rendered in real-time. While this system can make exercise more effective and more enjoyable for users, it does not actually motivate users to start exercising. The MOPET system was evaluated with 12 participants. During the user evaluation the MOPET system was compared with guiding users through fitness trails by written instructions in text. Results showed that the MOPET system was rated as more useful compared to written instructions and participants made fewer mistakes using the MOPET system [18]. One of the functions of the full body animated ECA in MOPET is to demonstrate the physical exercises, a function that should be clearly distinguished from its conversational functions.

There is a complexity of mediating factors that influences the user's experience of systems that use virtual characters to personify software agents. Experiments by Koda and Maes [40] already demonstrated that the perceived intelligence of a face is determined not by the agent's appearance but by its competence. There is a dichotomy between user groups which have opposite opinions about personification. Not all people are evenly inclined to anthropomorphize. Epley et al. [25] studies personal and situational factors that determine the need for socializing and the need to understand and control, the two factors that influence if someone anthropomorphizes. "Thus, agent-based interfaces should be flexible to support the diversity of users' preferences and the nature of tasks." [40]. A positive impact of features of an ECA on the perception of the user and on perceived credibility (by subjective measures) does not always come with a positive impact on long-term behavior change as the work of Bickmore and Picard [10] and Bickmore et al. [12] has shown.

In the next section we present the results of our own comparative field trials that contribute to more insight into the effects of the use of virtual humans as interface agents for personal coaching systems.

8.4   Kristina—Virtual Coach for Physical Activity

Office workers and many other adults in Western countries live a sedentary lifestyle. They sit too much and they don't reach the thirty minutes daily moderate intensive physical activity recommended by the World Health Organisation. Awareness of the risks involved and motivation to change one's lifestyle are the main conditions for a positive change. Wearable unobtrusive sensor technology has become cheaper and smaller over the last years. They can help people to at least monitor their physical activity level. Changing one's lifestyle is however easier said than done. A personal digital coach may help the user.

Kristina is a personal digital coaching system built to support and motivate users to live a balanced and healthy lifestyle. The system can target better adherence to medication intake and physical activity. To measure the amount of physical activity the system makes use of an activity monitor. To keep track of medication intake the system uses a digital pillbox. The coaching rules and the sensor data are stored on a central server that generates the feedback messages and send them to the device in use. Each output device has a set of available presentation formats and it is up to the user to select the format that he wants to use. On a smartphone a text message can be presented by simple text, it can be spoken or presented in a multi-modal way by an animated virtual human.

Kristina will motivate and support the users by providing information on the current behavior of the user (self-monitoring), by presenting feedback about the behavior of the users, such as the progress towards their personal goal. Hints how to reach these goal(s) will be presented to the users when appropriate. The messages and the devices that can be used to present feedback messages are based on user studies presented in Klaassen et al. [37]. Users have a preference for receiving feedback on the smartphone, computer, or television. Examples of situations in which users would mainly like to receive feedback in, is while relaxing in front of the television or while having a (lunch) break. As for the type of messages that can be sent, users seem to have a preference for reports about their progress, rather than advice or learning messages.

A coaching system can be real-time or off-line. In a real-time coaching system the sensor data is continuously available from the user for the coach who can decide to act anytime by sending feedback to the user. In some of these systems the data is sent through a Bluetooth connection to a mobile platform. Nowadays smartphones have built in activity sensors. op den Akker et al. [53] provides an overview of real-time

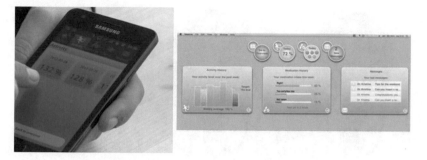

Fig. 8.2 Overview of physical activity presented on PC and smartphone

physical activity coaching systems. Kristina is not a real-time coaching system. The user has to connect the activity sensor physically to a PC in order to upload the data to the server. The user can only get feedback and a data overview on the data that was uploaded. This implies that in order for the coach to give timely feedback the user has to upload his data regularly, that is at least once every day. This is a disadvantage compared to real-time systems. In a real-time system the coach can send a motivational message saying "*It is time for a walk. You haven't been very active up to now today.*" Actually one of the tasks of Kristina is to remind the user to regularly upload his sensor data. Kristina uses strategies such as "tailoring", "suggestions" and "self-monitoring" and it targets users that are in the "action" state of the Transtheoretical Model (see Sect. 8.2.1 for more details about the Transtheoretical Model), i.e., they are motivated to change their lifestyle. Figure 8.2 gives an example of how data is presented to the user for self-monitoring.

8.4.1 Articulated Social Agent Platform

To take the full advantage of the power of embodied conversational agents in service systems it is important to support real-time, online and responsive interaction with the system through the embodied conversational agent. The Articulated Social Agent Platform (ASAP) is a model-based platform for the specification and animation of synchronised multi-modal responsive animated agents developed at the Universities of Twente and Bielefeld.[3] This is a realizer for the interpretation of embodied agent behavior specified in the Behavior Markup Language (BML) [67]. The back-end animator can be a full 3D graphical avatar and is running on desktop PC or a physical robot. Klaassen et al. [38] presents a light-weight PictureEngine as back-end engine which allows to run the platform in mobile Android applications. This makes it possible to generate real-time multi-modal behavior of an animated cartoon-like virtual human showing gestures and expressions, such as smiles and eye blink, synchronized

---

[3]http://asap-project.ewi.utwente.nl/ [64].

Fig. 8.3 Examples of different expressions of the virtual human used in the Kristina system. In this case dressed like a doctor

with synthetic speech. The ASAP platform is used in the Kristina personal coaching system. The user can choose how he will receive the messages of the coaching system through speech by an animated graphical character or by a text displayed on the screen. Figure 8.3 gives examples of the animated character. Here, Kristina is represented as a medical doctor. For speech we used the SVOX text-to-speech engine in English and Dutch for Android.

In the next section we will describe a user evaluation with the the mobile Kristina coaching in which we compare two versions one with and one without an animated graphical interface.

8.5   Coaching by Means of Text Messages or by a Virtual Human

In a long-term field study we investigated the possible effects of presenting Kristina on a smart phone by means of an animated virtual character. We measured effects on usability, user experience, likability, the quality of coaching, anthropomorphism, the credibility of the Kristina system and the usage of the system as well as the performance in terms of the user's physical activity. The three user groups that participated in this study are listed below. The evaluation was held in the period of January to August 2013.

- Text group—A group that used the coaching application on their smartphone and computer and received feedback messages in text.
- Virtual Human group—A group that used the coaching application on their smartphone and computer and received feedback presented by a virtual human.
- Control group—A group that used the coaching application on their computer and received feedback in text on the website, not on their smartphone.

Fig. 8.4 A feedback message presented in text and presented by a virtual human on a smartphone

8.5.1 Methodology

We used a between subjects design to evaluate the coaching system. The independent variables are the way in which feedback messages were presented to the users. Feedback messages could be presented in text or via a virtual human (see Fig. 8.4). Participants of the control group did not receive feedback messages actively and they did not receive notifications about new feedback messages. The dependent variables of this evaluation were the physical activity levels (PAL), the person's total energy expenditure (TEE) in a 24-hour period, divided by his or her basal metabolic rate [69], the number of uploads of physical activity data and the user perception. The dependent variables were measured by log data, questionnaires, and interviews.

Participants were asked to complete three questionnaires. The intake questionnaire was meant to collect background information like age, weight, height, gender and experiences with smartphones/PDAs, virtual humans and digital coaching systems. The second and third questionnaires (halfway through and at the end of the evaluation) were meant to evaluate the participants' experiences with the different feedback versions. We used questionnaires to measure the usability (System Usability Scale [15]), user experience (AttrakDiff2 [30]), quality of coaching of the coaching system (an adapted version of the Coaching Behaviour Scale for Sport (CBS-S) [23]), credibility (Source Credibility Scale [46]) and acceptance (UTAUT [65] and Godspeed [4]).

At the end of the user evaluation participants were invited to discuss their experiences in a semi-structured interview.

8.5.1.1 Subjects

Participants were asked to join the user evaluation by email, social media and face-to-face communication. Participants had to be office workers (sedentary profession) and should own their own Android smartphone with Android version 2.3 or higher. Sixty office workers indicated they were willing to join the experiment and were divided into the three groups. The distribution over the three groups was random. Participants in the virtual human group had to install the Dutch text-to-speech engine from the Google Play store.

8.5.1.2 Procedure

The duration of the complete evaluation was seven weeks. This included one assessment week at the start of the user evaluation. Before the start of the evaluation participants received the activity sensor and an information sheet with details about the user evaluation including an informed consent to be signed. Software, applications for the smartphone, user manuals about the installation of software and links to questionnaires were provided via email.

The first week was an assessment week in order to establish normal activity level for tailoring the system and to set a personal goal. During this assessment week no feedback messages were given by the system. After the assessment week participants received their personal goals and used the system for six weeks. In these six weeks feedback (updates, requests, reminders and overviews) about their progress was provided. Participants were asked to upload their activity data at least one time per day by connecting their activity monitor to their computer. Participants who over or under performed in the first three weeks were offered a new (higher or lower) goal by the system. Participants were free to accept this new goal or not.

Questionnaires were available online and participants received an email when it was time to fill in a questionnaire. Shortly after the evaluation the participant was visited again to collect the materials. During this visit participants were invited for a short final interview to discuss experiences.

8.5.2 Results

Forty-three participants completed the user evaluation by finishing the assessment week, using the system for at least six weeks and completing all the questionnaires. Participants were between 21 and 57 years old (average $= 36.5 \pm 10.4$), and worked 36.1 hours per week on average. All participants except two owned an Android smartphone. Those two participants were included in the control group. All the participants were familiar with mobile Internet. Most of the participants did sports, forty indicated doing sports every week.

Seventeen participants did not finish the user evaluation due to several personal reasons. Two participants (from the text group) indicated having difficulties in answering some of the questions from the questionnaires. They were excluded from the results of the questionnaires. Twenty-one participants participated in the interview at the end of the user evaluation study.

8.5.2.1 Usability

The System Usability Scale from Brooke [15] was used to measure the usability of the system. The system usability score can range from 0 to 100. We analysed the results of the questionnaires to investigate differences between the text group and the virtual human group by using an independent samples t-test. To investigate differences between the halfway and end questionnaires we used a paired samples t-test. The system usability score of the Kristina system was above 62.2, which can be seen as acceptable. The System Usability Scale was reliable (Cronbach's alpha = 0.716). From the results of the halfway questionnaire we found one significant difference on the statement "*I thought there was too much inconsistency in this system*"between the text group and the virtual human group (Group Text: $M = 3.65, SD = 0.70$, Group VH: $M = 2.93, SD = 1.16, t(30) = -2.131, p = 0.041$). From these results we can conclude that the virtual character of the Kristina coaching system made the system more inconsistent compared to the Kristina coaching system that was using text messages.

We asked the participants about their general impression of the coaching system. Four people mentioned explicitly that they liked obtaining awareness about the amount of physical activity. Participants tried to reach their daily goal and mentioned that they used the system in a different way over time. In the beginning they were more enthusiastic about the system compared to the end of the user evaluation.

We asked the participants about their experiences related to the feedback and overviews presented by the system. Remarks from the participants were about the timing and the content of the feedback messages. Feedback messages were seen as a kind of standard. Participants wanted to receive more detailed feedback. Participants stated that the system should become more intelligent or more exciting. Some suggest to use history data in feedback messages. The timing and content of the feedback messages were seen as predictable. Two participants mentioned that receiving feedback and/or tips was not always useful, because users do not always have the ability to follow the tips or reminders from the system. Three participants stated that feedback about physical activity was nice to have.

Participants were asked to give their opinion about the question whether such a system is useful to live a healthier life. From the text group eight participants answered this question with "Yes" and two participants answered with "No". Participants who answered "Yes" stated that the system created awareness, but feedback messages needed to be further developed to become more concrete. Participants who answered "No" stated that the system was only useful to live a healthier life for people who were already motivated to use such a system or the system should be advised by a doctor or health provider. Eight participants from the virtual human group also answered this question with "Yes" and two participants answered with "No". Participants who answered "Yes" stated that the system created awareness about physical activity, but it was important that the user was motivated to use the system and change their behavior. Participants who answered "No" stated that the timing and content of the feedback should be improved and the fact that the system will only help to lead a healthier lifestyle in the short term.

Table 8.1 Differences between the text and virtual human group on word pairs of the Attrakdiff questionnaire halfway through and at the end of the user evaluation on a 5 point Likert scale

Word pair	Text ($M(SD)$)	VH ($M(SD)$)	Independent samples test
Halfway			
"Unprofessional–professional"	4.94(1.09)	3.93(1.22)	$t(30) = 2.468$, $p = 0.020$
"Cautious–bold"	4.13(0.49)	3.67(0.90)	$t(30) = 1.795$, $p = 0.083$
End			
"Unruly–manageable"	5.36(1.17)	4.40(1.35)	$t(30) = -2.138$, $p = 0.041$

8.5.2.2 User Experience

User experience of the coaching system was measured using the AttrakDiff2 questionnaires. We analysed the results of the AttrakDiff2 questionnaires to investigate differences between the text group and the virtual human group by using independent samples t-test. To investigate differences between the halfway and end questionnaires we used a paired samples t-test. The reliability of all four dimensions was good (Cronbach's alpha > 0.6). Table 8.1 presents the significant results of the AttrakDiff2 questionnaires. From these results we can conclude that feedback messages presented by a virtual character were seen as less professional and less manageable compared to the same feedback messages presented in text.

8.5.2.3 Likeability

The acceptance of the coaching application was measured using a questionnaire based on the UTAUT questionnaire. We analysed the results of the questionnaires to investigate differences between the text group and the virtual human group by using an independent samples t-test. All constructs were reliable (Cronbach's alpha > 0.6) except for the anxiety (Anx) construct. From the results of the halfway questionnaire we found one marginally significant difference between the text group and the virtual human group on the statement "*The system would make my life more interesting*" (Group Text: $M = 1.71$, $SD = 1.10$, Group VH: $M = 2.47$, $SD = 0.52$, $t(30) = -1.927$, $p = 0.063$). This difference could not be found at the end of the user evaluation.

8.5.2.4 Anthropomorphism

The Godspeed questionnaire of Bartneck et al. [3] was used to measure the acceptance of the virtual coach. We analysed the results of the questionnaires to investigate

Table 8.2 Differences between the text and virtual human group on word pairs of the Godspeed questionnaire halfway through the user evaluation on a 5 point Likert scale

Word pair	Text ($M(SD)$)	VH ($M(SD)$)	Independent samples test
"Fake–natural"	2.34(0.86)	1.87(0.64)	$t(30) = 1.791$, $p = 0.083$
"Dead–alive"	2.71(0.69)	2.20(0.94)	$t(30) = 1.752$, $p = 0.090$
"Anxious–relaxed"	3.12(0.33)	3.47(0.74)	$t(30) = -1.751$, $p = 0.090$
"Still–surprised"	2.53(0.72)	2.07(0.80)	$t(30) = 1.727$, $p = 0.094$

Table 8.3 Differences between the text and virtual human group on word pairs of the Godspeed questionnaire at the end of the user evaluation on a 5 point Likert scale

Word pair	Text ($M(SD)$)	VH ($M(SD)$)	Independent samples test
"Unconscious–conscious"	2.71(0.69)	1.87(0.64)	$t(30) = 3.563$, $p = 0.001$
"Artificial–lifelike"	2.59(0.71)	1.80(0.68)	$t(30) = 3.199$, $p = 0.003$
"Dead–alive"	3.00(0.71)	2.33(0.82)	$t(30) = 2.476$, $p = 0.019$
"Unintelligent–intelligent"	3.012(0.49)	2.40(1.12)	$t(30) = 2.401$, $p = 0.023$
"Fake–natural"	2.65(0.79)	2.13(0.83)	$t(30) = 1.793$, $p = 0.083$
"Machinelike–humanlike"	2.59(0.94)	2.00(0.93)	$t(30) = 1.789$, $p = 0.085$
"Still–surprised"	2.59(0.62)	2.13(0.74)	$t(30) = 1.890$, $p = 0.068$

differences between the text group and the virtual human group by using an independent samples t-test. All constructs were reliable (Cronbach's alpha > 0.6) except for the perceived safety construct. Differences between the text group and the virtual human group halfway the user evaluation can be found in Table 8.2. Differences between these groups at the end of the user evaluation can be found in Table 8.3.

The coaching system that was using the virtual character to present feedback messages was seen as more artificial, fake, dead, and machine-like. The system using the virtual character was also seen as less intelligent compared to the system that was using text messages.

Participants were asked for their opinion on the two presentation forms on the smartphone. Before that we gave a short explanation and demonstration of both the virtual human coach and the text coach on the smartphone of the researcher. Three participants from the text group stated that a virtual coach represented by text was fine for them because they were afraid that a virtual human would become annoying or they did not like the idea of having a virtual human as an user interface telling them what to do. Four participants from the text group stated that it would not make any difference in this coaching system. Participants stated that the system would stay mechanical and predictable because it would not change the behavior of the complete system. Some stated that the motivation to use such a system was more important than the way that a coach is represented. Three participants from the text group had the impression that a coach represented by a virtual human was more fun to use, more human-like or closer to the way a normal coach would provide feedback. Eight of the participants from the virtual human group stated that they preferred feedback messages in text. They had the impression that the virtual human did not add anything extra compared to a text message for the kind of feedback messages presented in this coaching system. Five of these participants stated that the virtual human can be useful when more detailed feedback will be presented, when it is possible to have interaction with the coach or when users need extra motivation. Two participants preferred feedback presented by the virtual human. In their opinion the virtual human added some extra dimension to the coaching system, it was fun to get feedback presented by a virtual human and it added some social pressure to the feedback message.

8.5.2.5   Quality of Coaching

We used an adapted version of the CBS-S to measure the quality of coaching by the system in the different conditions. We analysed the results of the questionnaires to investigate differences between the text group and the virtual human group by using independent samples t-test. The reliability is good (Cronbach's alpha > 0.958). There was no difference between virtual human and text group in the perceived quality of coaching.

8.5.2.6   Credibility

The credibility of the feedback messages and the coaching system was measured using the Source Credibility Scale of McCroskey and Young [48]. To investigate differences between the halfway and end questionnaires we used a paired samples t-test. The reliability was good (Cronbach's alpha > 0.88). Results of the Source Credibility Scale can be found in Table 8.4.

From the results we can conclude that the Kristina coaching system that was using the virtual character was seen as less intelligent compared to the system that was using text messages. Feedback messages presented by the virtual character were seen as more honest.

Table 8.4 Differences between the text and virtual human group on the Source Credibility Scale halfway through and at the end of the user evaluation on a 5 point Likert scale

Word pair	Text ($M(SD)$)	VH ($M(SD)$)	Independent samples test
Halfway			
"Unintelligent–intelligent"	4.35(1.11)	3.60(1.12)	$t(30) = -1.902$, $p = 0.084$
End			
"Dishonest–honest"	4.53(1.23)	5.33(1.11)	$t(30) = 1.928$, $p = 0.063$
"Unselfish–selfish"	4.47(0.87)	5.07(1.10)	$t(30) = 1.702$, $p = 0.098$
"Sinful–virtuous"	4.06(0.43)	4.60(0.99)	$t(30) = 2.057$, $p = 0.048$

8.5.3   Log Data

During the user evaluation data about the physical activity levels (PAL), the number of uploads of activity data and the number of feedback messages sent by the coach were collected.

Activity data

Figure 8.5 presents the physical activity level of all the participants per group. We analysed the physical activity levels of each participant and used a Mixed Models

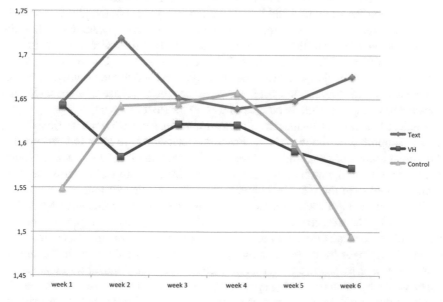

Fig. 8.5 Physical activity levels per week. The lines represent the average PAL per group

approach to investigate whether time (in weeks) or group (condition) was a predictor for the level of physical activity. From this test we can conclude that there was no relationship between time or condition and the physical activity level. We used a Mann-Whitney test to compare the average values of all participants of the groups to investigate whether there is a difference between the groups. At the end of week 6 the PAL level of the control group was significantly less than the text group ($Mdn = 1.61$, $U = 36.00$, $r = 0.460$, $p = 0.014$).

Number of uploads and feedback messages

Participants were asked to upload their physical activity data on a regular basis, preferable once a day. During the six weeks (42 days) user evaluation participants uploaded their activity data 32 times on average. We used a Mann-Whitney test to compare the results between the different groups. The number of uploads of physical activity data in the control group was significantly less than in the Text group during all six weeks ($Mdn = 3.00$, $U = 51.00$, $p = 0.045$, $r = 0.378$). No difference was found between the text group and virtual human group. Over the whole period on average 74 feedback messages were sent per participant.

8.5.4   Conclusion

We can conclude that presenting feedback messages in text was a better approach compared to presenting feedback messages by a virtual human. In general the feedback messages were seen as a kind of standard and participants stated that there was too much repetition. Behavioral measures (log data from sensor) showed no differences between the text group and the virtual human group, but the physical activity level of the Text group was increasing in week 6 of the user evaluation, while the physical activity levels of the virtual human and control group were decreasing. We found a significant difference between the Text group and the Control group at the end of the user evaluation when we look at the physical activity level and the number of uploads of data. Regular feedback and reminders help but use of a virtual human does not have a positive effect compared to a simple text message.

Subjective measures obtained by the questionnaires were in favor of the text coach. This is supported by the interviews. From the interviews we have indications that the virtual human coach could improve users' perception if the feedback messages become more intelligent, if interaction with the virtual coach was possible or when users are motivated more to use a physical activity coaching system.

Perceived quality of a system is a subjective notion. It is influenced by desires as well as by the expectations users have [68]. Participants from the two groups that received feedback messages had different reference points to which they compared the messages from the coaching system. Using a virtual human in a coaching application that makes use of verbal and non-verbal communication creates high expectations by the users. Using speech output may create the expectation that the virtual human is able to recognise verbal input, a feature not implemented in Kristina.

The context of use influences the acceptance of the verbal animated presentation of feedback messages and hints by the virtual coach. Short attention spans are typical in mobile use [34]. In the mobile context, many other tasks and other applications on smartphones ask for the users' attention. Text messages are easier and quicker to read and understand compared to a message presented by a virtual human that uses text to speech thereby controlling the timing and duration of the presentation of the message. Glanceablility of the interface affects the perceived ease of use and usefulness of the system and thereby the efficacy of the coaching [14].

8.6 Guidelines for Use of Virtual Coaches

We have seen that a combination of properties of the user (motivation, needs and expectations), the (mobile) use context, the task of the virtual coach (to give short hints and feedback or to motivate) and the appearance of a virtual coach determine the way users experience and react to personal coaching systems.

Based on lessons learned from our experiences with the Kristina system and the literature reviews we can now present some guidelines and considerations for deciding when to use an embodied conversational agent in the interface of a personal coaching system.

1. Utilitarian functions come first, hedonic aspects are important but come second place.
2. Context of use is of prime importance and impacts attention spans and communication efficiency.
3. Dependent on the task of the system and the target user group it is sometimes better not to use a virtual human that represents an authority like a personal trainer. An animated emphatic character should be considered.
4. The functions of the (animated) graphical interface of a persuasive system should be reflected in a coherent way in appearance, expressions and behaviors and be related to the utilitarian functions of the system.
5. Do not use verbose text messages when simple signals can transmit the same message.
6. The user must have control over the system and the way user data is handled.
7. Given the state of the art of spoken dialog systems use of a spoken dialog may have negative impact on the user acceptance if not handled with care.
8. Do not let the system give information or suggest or advice actions based on uncertain information.
9. Let the system be clear about the intention and the cognitive status of information it provides to the user.
10. Make your persuasive actor so that it is aware of the fact that some users don't like to be bothered by calls for attention they didn't ask for.
11. Be aware of the fact that as designer you are responsible for making the system so that it allows the user to act in a responsible way.

We believe there is potential in the use of virtual humans for personal coaching systems when used in the appropriate way. The quality of automatic speech recognition and speech generation systems has improved so that interaction between the user and the system by means of a natural conversation comes within reach. Speech interfaces have arrived in every day use. In 2011 Apple introduced their speech enabled personal assistant Siri. Android mobile users can now choose from a wealth of similar personal assistants: Google Now, Voice Actions, Indigo, and many more. These commercial speech interfaces do not allow real dialogues. The interaction consists of isolated question response pairs. Spoken dialogue systems is an active research field adding to the development of embodied conversation agents that allow a conversational style of human computer interaction [42]. Progress in the quality of speech recognition and the use of more sophisticated spoken dialog systems designed for this type of application domains will improve the quality of the interaction and create new potentials for interaction between users and virtual coaches through the use of embodied conversational agents. Dialogs can either be used to collect information from the user that is hard or not to obtain through sensors, for motivational interviewing, or for persuasive argumentation [1, 28].

Future will show what role personal coaching systems can play in cooperation with patients and medical experts in clinical settings and how we can best exploit the strong points of computer systems on the one hand and those of the human on the other hand. Care for our selves, our environment and others cannot be transmitted to technical systems. Despite the fact that we sometimes perceive them as caring social actors.

Acknowledgments Thanks to all Smarcovians who contributed in one way or another to the work reported here. This work was funded by the European Commission, within the framework of the ARTEMIS JU SP8 SMARCOS project 100249

References

1. Andrews P, Manandhar S, Boni MD (2008) Argumentative human computer dialogue for automated persuasion. In: Proceedings of 9th SIGdial workshop on discourse and dialogue, association for computational linguistics, pp 138–147
2. Bandura A (1977) Self-efficacy: toward a unifying theory of behavioral change. Psychol Rev 84(2):191–215
3. Bartneck C, Croft E, Kulic D (2008) Measuring the anthropomorphism, animacy, likeability, perceived intelligence and perceived safety of robots. In: Metrics for HRI workshop, technical report, vol 471. Citeseer pp 37–44
4. Bartneck C, Kuli D, Croft E, Zoghbi S (2009) Measurement instruments for the anthropomorphism, animacy, likeability, perceived intelligence, and perceived safety of robots. Int J Soc Robot 1:71–81
5. Baylor AL (2011) The design of motivational agents and avatars. Educ Technol Res Dev 59(2):291–300
6. Beale R, Creed C (2009) Affective interaction: how emotional agents affect users. Int J Hum-Comput Stud 67(9):755–776

7. Berdichevsky D, Neuenschwander E (1999) Toward an ethics of persuasive technology. Commun ACM 42(5):51–58
8. Berry DC, Butler LT, de Rosis F (2005) Evaluating a realistic agent in an advice-giving task. Int J Hum-Comput Stud 63(3):304–327
9. Bettinghaus EP, Cody MJ (1994) Persuasive communication, 5th edn. Wadsworth, Ted Buchholz
10. Bickmore T, Picard R (2005) Establishing and maintaining long-term human computer relationships. ACM Trans Comput-Hum Interact 12:293–327
11. Bickmore T, Mauer D, Crespo F, Brown T (2007) Persuasion, task interruption and health regimen adherence. Proceedings of the 2nd international conference on persuasive technology (PERSUASIVE'07). Springer, Berlin, pp 1–11
12. Bickmore TW, Silliman RA, Nelson K, Cheng DM, Winter M, Henault L, Paasche-Orlow MK (2013) A randomized controlled trial of an automated exercise coach for older adults. J Am Geriatr Soc 61(10):1676–1683. doi:10.1111/jgs.12449
13. Blascovich J, Loomis J, Beall AC, Swinth KR, Hoyt CL, Bailenson JN (2002) Immersive virtual environment technology as a methodological tool for social psychology. Psychol Inq 13(2):103124
14. Boerema S, Klaassen R, op den Akker HJA, Hermens HJ (2012) Glanceability evaluation of a physical activity feedback system for office workers. In: Proceedings of EHST 2012: the 6th International symposium on ehealth services and technologies, SciTePress–science and technology publications, pp 52–57
15. Brooke J (1996) Sus-a quick and dirty usability scale. Usability Eval Ind 189:194
16. Burgoon JK, Birk T, Pfau M (1990) Nonverbal behaviors, persuasion, and credibility. Hum Commun Res 17(1):140–169
17. Buttussi F, Chittaro L (2008) Mopet: a context-aware and user-adaptive wearable system for fitness training. Artif Intell Med 42(2):153–163
18. Buttussi F, Chittaro L, Nadalutti D (2006) Bringing mobile guides and fitness activities together: a solution based on an embodied virtual trainer. Proceedings of the 8th conference on human-computer interaction with mobile devices and services (MobileHCI '06). ACM, New York, pp 29–36
19. Cassell J (2001) Embodied conversational agents: representation and intelligence in user interfaces. AI Mag 22(4):67–83
20. Cassell J, Pelachaud C, Badler N, Steedman M, Achorn B, Becket T, Douville B, Prevost S, Stone M (1994) Animated conversation: rule-based generation of facial expression, gesture and spoken intonation for multiple conversational agents. Proceedings of the 21st annual conference on Computer graphics and interactive techniques (SIGGRAPH '94). ACM, New York, pp 413–420
21. Cassell J, Sullivan J, Prevost S, Churchill EF (eds) (2000) Embodied conversational agents. MIT Press, Cambridge
22. Catrambone R, Stasko J, Xiao J (2002) Anthropomorphic agents as a user interface paradigm: Experimental findings and a framework for research. In: Proceedings of the 24th annual conference of the cognitive science society, pp 166–171
23. Cote J, Yardley J, Hay J, Sedgwick W, Baker J (1999) An exploratory examination of the coaching behaviour scale for sport. AVANTE 5:82–92
24. Dehn D, van Mulken S (2000) The impact of animated interface agents: a review of empirical research. Int J Hum-Comput Stud 52(1):1–22
25. Epley N, Waytz A, Akalis S, Cacioppo J (2008) When we need a human: motivational determinants of anthropomorphism. Soc Cognit 26(2):143–155
26. Fogg B (2002) Persuasive technology: using computers to change what we think and do. Morgan Kaufmann Publishers, San Francisco
27. Fogg B (2009) A behavior model for persuasive design. In: Proceedings of the 4th international conference on persuasive technology (Persuasive '09). ACM, New York, pp 40:1–40:7
28. Grasso F (2003) Rhetorical coding of health promotion dialogues. In: Dojat M, Keravnou E, Barahona P (eds) Artificial intelligence in medicine, vol 2780. Lecture notes in computer science. Springer, Berlin, pp 179–188

29. Guadagno RE, Blascovich J, Bailenson JN, McCall C (2007) Virtual humans and persuasion: the effects of agency and behavioral realism. Media Psychol 10:122
30. Hassenzahl M (2008) User experience (UX): towards an experiential perspective on product quality. Proceedings of the 20th international conference of the association francophone d'Interaction homme-machine (IHM '08). ACM, New York, pp 11–15
31. Heckman CE, Wobbrock JO (2000) Put your best face forward: anthropomorphic agents, e-commerce consumers, and the law. Proceedings of the fourth international conference on autonomous agents (AGENTS '00). ACM, New York, pp 435–442
32. Henkemans OAB, van der Boog P, Lindenberg J, van der Mast C, Neerincx M, Zwetsloot-Schonk BJHM (2009) An online lifestyle diary with a persuasive computer assistant providing feedback on self-management. Technol Health Care 17:253–257
33. Hermens H, op den Akker H, Tabak M, Wijsman J, Vollenbroek M (2014) Personalized coaching systems to support healthy behavior in people with chronic conditions. J Electromyogr Kinesiol 24(6):815–826
34. Kaasinen E, Roto V, Roloff K, Väänänen-Vainio-Mattila K, Vainio T, Maehr W, Joshi D, Shrestha S (2009) User experience of mobile internet analysis and recommendations. J Mob HCI Spec Issue Mob Internet User exp 1(4):4–23
35. Kay A (1990) User interface: a personal view. In: Laurel B (ed) The art of human-computer interface design. Addison-Wesley, Reading, pp 191–208
36. Klaassen R, op den Akker R, op den Akker H (2013) Feedback presentation for mobile personalised digital physical activity coaching platforms. In: Proceedings of the 6th international conference on pervasive technologies related to assistive environments (PETRA 2013). ACM, New York, p 8
37. Klaassen R, op den Akker R, Lavrysen T, Wissen S (2013) User preferences for multi-device context-aware feedback in a digital coaching system. J Multimodal User Interfaces 21:1–21
38. Klaassen R, Hendrix J, Reidsma D, op den Akker R, van Dijk B, op de Akker H (2013) Elckerlyc goes mobile–enabling natural interaction in mobile user interfaces. Int J Adv Telecommun 6(1–2):45–56
39. Klasnja P, Consolvo S, Pratt W (2011) How to evaluate technologies for health behavior change in HCI research. Proceedings of the SIGCHI conference on human factors in computing systems (CHI '11). ACM, New York, pp 3063–3072
40. Koda T, Maes P (1996) Agents with faces: the effect of personification. In: Proceedings of 5th IEEE international workshop on robot and human communication, pp 189–194
41. Kozierok R, Maes P (1993) A learning interface. Proc AAAI 1993:459–465
42. Lemon O (2012) Conversational interfaces. In: Lemon O, Pietquin O (eds) Data-driven methods for adaptive spoken dialogue systems. Springer, Berlin, pp 1–4
43. Lester JC, Converse SA, Kahler SE, Barlow ST, Stone BA, Bhogal RS (1997) The persona effect: affective impact of animated pedagogical agents. Proceedings of the SIGCHI conference on human factors in computing systems (CHI '97). ACM, New York, pp 359–366
44. Li J (2015) The benefit of being physically present: a survey of experimental works comparing copresent robots, telepresent robots and virtual agents. Int J Hum-Comput Stud 77:23–37
45. Mazzotta I, Novielli N, De Carolis B (2009) Are ECAs more persuasive than textual messages? In: Ruttkay Z, Kipp M, Nijholt A, Vilhjálmsson HH (eds) Intelligent virtual agents, vol 5773. Lecture notes in computer science. Springer, Berlin, pp 527–528
46. McCroskey J (1966) Scales for the measurement of ethos. Speech Monogr 33(1):65–72
47. McCroskey J (1967) The effects of evidence in persuasive communication. West Speech 31:189–199
48. McCroskey JC, Young TJ (1981) Ethos and credibility: the construct and its measurement after three decades. Commun Stud 32(1):24–34
49. Murano P (2006) Why anthropomorphic user interface feedback can be effective and preferred by users. In: Chen CS, Filipe J, Seruca I, Cordeiro J (eds) Enterp Inf Syst VII. Springer, Netherlands, pp 241–248
50. Murano P, Gee A, Holt PO (2011) Evaluation of an anthropomorphic user interface in a travel reservation context and affordances. J Comput 3:8

51. Oinas-Kukkonen H, Harjumaa M (2008) A systematic framework for designing and evaluating persuasive systems. Proceedings of the 3rd international conference on persuasive technology (PERSUASIVE '08). Springer, Berlin, pp 164–176

52. op den Akker R, Klaassen R, Lavrysen T, Geleijnse G, van Halteren A, Schwietert H, van der Hout M (2011) A personal context-aware multi-device coaching service that supports a healthy lifestyle. Proceedings of the 25th BCS conference on human-computer interaction (BCS-HCI '11). British Computer Society, Swinton, pp 443–448

53. op den Akker H, Jones VM, Hermens HJ (2014) Tailoring real-time physical activity coaching systems: a literature survey and model. User Model User-Adapt Interact 24:351–392

54. op den Akker H, Cabrita M, op den Akker R, Jones VM, Hermens HJ (2015) Tailored motivational message generation: a model and practical framework for real-time physical activity coaching. J Biomed Inform 55:104–115

55. Petty R, Cacioppo J (1981) Attitudes and persuasion: classic and contemporary approaches. William C, Brown, Dubugue

56. Petty R, Cacioppo J (1986) Communication and persuasion: central and peripheral routes to attitude change. Springer, Berlin

57. Prochaska JO, Velicer WF, Rossi JS, Goldstein MG, Marcus BH, Rakowski W, Fiore C, Harlow LL, Redding CA, Rosenbloom D, Rossi SR (1994) Stages of change and decisional balance for twelve problem behaviors. Health Psychol 13:39–46

58. Reeves B, Nass C (1996) The media equation: how people treat computers, television, and new media like real people and places. Cambridge University Press, Cambridge

59. Reidsma D, Ruttkay Z, Nijholt A (2007) Challenges for virtual humans in human computing. In: AI for human computing. Lecture notes in artificial intelligence, vol 4451. Springer, Berlin, pp 316–338

60. Schulman D, Bickmore TW (2009) Persuading users through counseling dialogue with a conversational agent. Persuasive '09: proceedings of the 4th international conference on persuasive technology. ACM, New York, pp 1–8

61. Shneiderman B, Maes P (1997) Direct manipulation versus interface agents. Interactions 4(6):42–61

62. Sproull L, Subramani M, Kiesler S, Walker JH, Waters K (1996) When the interface is a face. Hum Comput Interact 11(2):97–124

63. Turunen M, Hakulinen J, Stahl O, Gamback B, Hansen P, Gancedo MCR, de la Cmara RS, Smith C, Charlton D, Cavazza M (2011) Multimodal and mobile conversational health and fitness companions. Comput Speech Lang 25(2):192–209

64. van Welbergen H, Yaghoubzadeh R, Kopp S (2014) AsapRealizer 2.0: the next steps in fluent behavior realization for ECAs. In: Intelligent virtual agents. LNCS, vol 8637. Springer, Berlin, pp 449–462

65. Venkatesh V, Morris MG, Davis GB, Davis FD (2003) User acceptance of information technology: toward a unified view. MIS Q 27:425–478

66. Verbeck PP (2006) Persuasive technology and moral responsibility: toward an ethical framework for persuasive technologies. Persuasive 6:1–15

67. Vilhjalmsson H, Cantelmo N, Cassell J, Chafai NE, Kipp M, Kopp S, Mancini M, Marsella S, Marshall AN, Pelachaud C, Ruttkay Z, Thorisson KR, van Welbergen H, van der Werf RJ (2007) The behavior markup language: recent developments and challenges. In: Pelachaud C, Martin JC, Andre E, Collet G, Karpouzis K, Pele D (eds) Intelligent virtual agents. Lecture notes in computer science, vol 4722. Springer, Berlin, pp 99–111

68. Wechsung I (2013) An evaluation framework for multimodal interaction: determining quality aspects and modality choice. Springer, Berlin

69. Westerterp KR (1999) Obesity and physical activity. Int J Obes 23(1):59–64

70. Yee N, Bailenson J, Rickertsen K (2007) A meta-analysis of the impact of the inclusion and realism of human-like faces on user experiences in interfaces. Proceedings of the SIGCHI conference on human factors in computing systems (CHI '07). ACM, New York, pp 1–10

Chapter 9
Social Perception in Machines: The Case of Personality and the Big-Five Traits

Alessandro Vinciarelli

Abstract Research agendas aimed at the development of socially believable behaving systems often indicate automatic social perception as one of the steps. However, the exact meaning of the word "perception" seems still to be unclear in the computing community, in particular when it applies to social and psychological phenomena that are not accessible to direct observation. This chapter tries to shed light on the problem by showing examples of approaches that perform Automatic Personality Perception, i.e. the prediction of personality traits that people attribute to others.

9.1 Introduction

Endowing machines with social perception is one of those steps that many research agendas indicate as crucial towards the development of socially believable behaving systems (see, e.g., [15, 26]). However, the exact meaning of the expression "*perception*" is still unclear in the computing community, especially when it comes to social and psychological aspects. Furthermore, there are peculiar aspects of the experiments in the field (e.g., the low agreement between raters) that still attract concerns while they should be considered an inherent characteristic of the social perception problem (see, e.g., [10, 11] for the case of personality traits). This chapter tries to address the issue, at least to a certain extent, by providing a simple conceptual model of social perception and by showing a few examples related to Automatic Personality Perception (APP) [24], the task of predicting how people perceive the personality of others.

Social perception is the inference of socially relevant information about others, including their intentions, emotions, beliefs, traits, inner state, etc. Social cognition investigations have shown that the phenomenon has two main characteristics:

A. Vinciarelli (✉)
School of Computing Science and Institute of Neuroscience and Psychology,
Universiy of Glasgow, Glasgow, UK
e-mail: Alessandro.Vinciarelli@glasgow.ac.uk

© Springer International Publishing Switzerland 2016
A. Esposito and L.C. Jain (eds.), *Toward Robotic Socially Believable
Behaving Systems - Volume II*, Intelligent Systems Reference Library 106,
DOI 10.1007/978-3-319-31053-4_9

The first is that it is spontaneous (*"People make social inferences without inten-tions, awareness or effort, i.e., spontaneously"* [22]), and the second is that it is fast (*"Within milliseconds of encountering others, we readily process and react to infor-mation about their various personal characteristics and social group memberships [...]"* [9]).

One of the main consequences of the characteristics above is that social infer-ence processes are, at least to a certain extent, mechanic: *"it now appears that when we observe a person behave, we jump automatically to conclusions about the per-son's underlying traits [...]"* [12]. In other words, the relationship between stimuli (everything observable people do) and outcomes of the social perception processes (attributed goals, emotions, traits, inner state, etc.) is likely to be stable enough to be modeled in statistical terms. This is a major advantage for the development of auto-matic social perception processes because computing science, in particular machine learning, provides a wide spectrum of statistical methods aimed at modeling statisti-cal relationships like those observed in social perception. Hence, at least in principle, it should be possible to reproduce human social perception processes with machines.

The main reason for addressing the social perception problem is that people behave towards others according to the social inferences they make, especially in the earliest stages of an interaction [23]. Therefore, reproducing social perception processes in machines will help to achieve the goal of socially believable behaving agents, i.e. agents that interact with their users like humans do. However, there are other technologies that can benefit from these efforts as well. One is the inference of how people react to individuals they hear in audio recordings (e.g., through the radio) or see in videos (e.g., they see on TV). This has been shown, e.g., to explain how people vote according to the appearance of candidates [21] or what traits social media users attribute to people that tag certain images as favorite [25]. Another relevant field is Human-Computer Interaction, in particular for what concerns the tendency of people to attribute human characteristics to machines even when these do not display human-like behaviors [14]. More in general, any application involving the interaction with humans might benefit from technologies revolving around social perception.

The rest of this paper is organized as follows: Sect. 9.2 proposes a simple con-ceptual model of how social perception works, Sect. 9.3 proposes examples based on Automatic Personality Perception, and Sect. 9.4 draws some conclusions.

9.2 A Conceptual Model

Figure 9.1 shows a simplified version of the Brunswik's Lens [4], one of the most widely accepted models of how living beings gather information in their environment. In the particular case of social perception, the Lens is a good conceptual model of how people infer socially relevant information about others.

9.2.1   Psychological Level

The scheme of Fig. 9.1 must be read from left to right: the *target* (the person on the left hand side) is characterized by a state that can correspond to any type of socially and psychologically relevant information (emotions, goals, attitude, etc.). The state is not directly observable, but gets manifested through an *externalization* process. This latter results into *distal cues*, i.e. cues accessible to the senses of a potential *observer* (the person on the right hand side). The sensory apparatus of the observer processes the cues and produces a *percept* characterized by *proximal cues*, i.e. the cues actually available to the cognition of the observer. The proximal cues activate an *attribution* process that result into a *perceived state*, i.e. the state that the observer infers from the proximal cues.

   The perceived state is not necessarily accurate, i.e. it does not necessarily correspond to the actual state of the target (this is why actual and perceived state have different shape in Fig. 9.1). This embodies the everyday intuition that people do not necessarily attribute to others the right intentions or do not understand how others actually feel. The potential disalignment between actual and perceived state explains why social perception studies are often characterized by low accuracy and low agreement between observers (see, e.g., [10, 11] for the Big-Five Traits).

9.2.2   Technical Level

The different steps of the Brunswik Lens correspond to the different stages of social perception approaches presented in the literature. Given that machines can work only

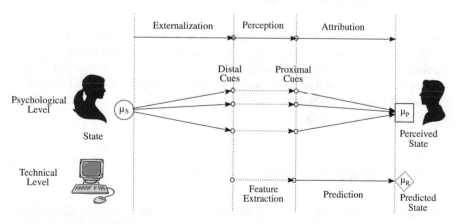

Fig. 9.1 The picture shows a simplified version of the Brunswik Lens. The psychological level describes the main stages of social perception. The technical level describes the main stages of automatic social perception approaches

on physical, machine detectable information, automatic approaches typically start with the detection of the distal cues or, in general, from the extraction of physical measurements—the *features*—from signals collected with sensors that capture the behavior of the target (e.g., speech signals or videos recorded with microphones and cameras, respectively).

Given that proximal cues are not accessible to sensors, automatic approaches for social perception typically use distal cues as an approximation. In principle, it is possible to model the process that leads from distal to proximal cues. However, to the best of our knowledge, no approach presented in the literature tries to do it. The physical measurements are typically arranged into vectors (the *feature vectors*) that can be fed to statistical inference approaches. These latter map the vectors into target values typically corresponding to the judgments made by raters that have observed the data. In general the judgments are obtained by using psychometric questionnaires that measure in quantitative terms a social or psychological phenomenon.

In general, the performance of the automatic approaches is measured in terms of how close are the outcomes of the approach to the judgments made by the raters. The main assumption behind such an evaluation approach is that social perception technologies are not expected to predict the actual state of the target, but the state that observers attributed to the target. This is an important difference with respect to traditional machine intelligence approaches that typically try to recognize a *groundtruth*, i.e. a truth value that can be measured in objective and unambiguous terms. However, this does not make a difference from a methodological point of view and all the techniques developed in fields like machine learning and pattern recognition can be applied with success to automatic social perception as well.

9.3 The Personality Example

The goal of this section is to show examples of how social perception approaches can actually be implemented. In particular, the section focuses on approaches aimed at predicting the traits that observers attribute to targets on the basis of speech and facial appearance, a task known as Automatic Personality Perception [24]. In terms of the Brunswik's Lens, the traits of a target are the actual state while the traits attributed by the observers are the attributed state.

9.3.1 Personality and Its Measurement

Personality is a psychological construct aimed at capturing *"individuals' characteristic patterns of thought, emotion, and behavior together with the psychological mechanisms—hidden or not—behind those patterns"* [5]. In other words, personality should account for stable individual characteristics that make an individual behave similarly, at least to a certain extent, in similar situations. The experiments presented in this chapter adopt the Big-Five (BF) Traits—five major dimensions known to account for most individual differences [19]—as a personality model. Besides being the most effective and widely accepted personality model [27], the BF has the advantage of representing personalities as five-dimensional vectors, a format particularly suitable for computer processing. Thus, it is not surprising to observe that almost all computing approaches dealing with personality adopt the BF [24].

The components of the vectors are scores that account for how well the tendencies associated to a trait describe the behavior of an individual:

- *Openness*: tendency to be intellectually open, curious and have wide interests;
- *Conscientiousness*: tendency to be responsible, reliable and trustworthy;
- *Extraversion*: tendency to interact and spend time with others;
- *Agreeableness*: tendency to be kind, generous, etc.;
- *Neuroticism*: tendency to experience the negative aspects of life, to be anxious, sensitive, etc.

The scores are obtained by filling personality questionnaires. These are lists of items associated to Likert scales where the points can be mapped into numbers (e.g., *strongly disagree* corresponds to −2 while *strongly agree* corresponds to 2). The answers to individual items can then be combined through sums and subtractions to provide the five scores corresponding to the five traits. Table 9.1 shows the *Big-Five Inventory 10* (BFI-10) [16], the personality questionnaire adopted for the experiments of this chapter. The main reason behind the choice is that the BFI-10 requires less than one minute to be filled (it includes only 10 items) and it is then particularly suitable for the collection of large corpora. However, the literature proposes a large number of other instruments (see [24] for a short survey).

Table 9.1 The BFI-10 [16] is the short version of the Big-Five Inventory

Item	Sign	Trait
This person is reserved	−	Ext
This person is generally trusting	+	Agr
This person tends to be lazy	−	Con
This person is relaxed, handles stress well	−	Neu
This person has few artistic interests	−	Ope
This person is outgoing, sociable	+	Ext
This person tends to find fault with others	−	Agr
This person does a thorough job	+	Con
This person gets nervous easily	+	Neu
This person has an active imagination	+	Ope

Each item is associated to a Likert scale, from −2 (*"Strongly disagree"*) to 2 (*"Strongly agree"*) and contributes to the integer score (in the interval [−4, 4]) of a particular trait, see the third column. The answers are mapped into numbers (e.g., from −2 to 2)

Table 9.2 The table shows the main details of the two corpora used in the examples presented in this chapter

Corpus	Speaker Personality Corpus	Face Personality Corpus
Samples	640	829
Length	1h:46m	N/A
No. of targets	322	829
Gender balance	78.5 % M/21.5 % F	57.0 % M/43.0 % F
Subject distribution	80 % <3	100 % =1
No. of observers	11	11

In the SSPNet Speaker Personality Corpus, 80 % of the subjects are represented less than 3 times, while in the SSPNet Face Personality Corpus all subjects are represented once

9.3.2 Data Collection

Social perception experiments focus on attributed traits (see Fig. 9.1). Therefore corpora for the experiments include two main elements, namely recordings portraying the targets (e.g., audio samples, videos or pictures) and judgments made by a pool of observers based on the recordings. Every observer must provide a judgment (a questionnaire like the BFI-10 in the case of personality) for all the recordings of the corpus. This typically limits the number of subjects in corpora for social perception experiments. In the case of personality, the largest corpora includes a few hundreds of subjects and involve a pool of $10-15$ observers.

As an example, it is possible to consider the *SSPNet Speaker Personality Corpus*[1] [13] and the *SSPNet Face Personality Corpus* [1]. The former is a collection of 640 speech samples assessed by 11 observers, the latter is a corpus of 829 face images assessed by 11 observers (the main details of the two corpora are available in Table 9.2).

9.3.3 How Many Observers? How Much Agreement?

One of the main peculiarities of social perception studies is that, unlike in other problems, observers are not expected to be necessarily accurate. Figure 9.1 shows that actual and attributed state are not necessarily the same. In the case of personality, this means that actual and attributed traits do not necessarily match each other. Still, this leaves open the problem of how many raters should be involved in an experiment and how much they should agree with one another.

The number of observers can be set using the *effective reliability* ρ [18]:

$$\rho = \frac{Nr}{1 + (N - 1)r}, \tag{9.1}$$

---

[1]http://sspnet.eu/2013/10/sspnet-speaker-personality-corpus/.

where r is the average inter-observer correlation and N is the total number of observers. The number N should be such that ρ reaches satisfactory levels (more than 0.8 according to a commonly applied thumb rule). Hence, a low number of observers is acceptable as long as r is sufficiently high while a large number of observers is needed when r is low.

When it comes to observer consensus, acceptable values depend on the particular inference being investigated. When it comes to the Big-Five, the literature suggests that the agreement should be measured in terms of amount of variance α shared by the observers [10, 11]. The measurement can be done with a two-way Analysis of Variance where the observers correspond to the treatments [6]:

$$\alpha = \frac{\sigma_s^2}{\sigma_s^2 + \sigma_r^2 + \sigma_{s\times r}^2},\tag{9.2}$$

where $\sigma_s^2 = (\mu_s - \mu_{s\times r})/R$ is the estimate of the subject variance, $\sigma_r^2 = (\mu_r - \mu_{s\times r})/S$ is the estimate of the raters' variance and $\sigma_{s\times r}^2 = \mu_{s\times r}$ is the estimate of the variance of the interaction between raters and subjects.

According to an extensive survey of agreement values in Big-Five perception [11], the median of the α values over the works presented in the literature is 0.07 for Openness, 0.13 for Conscientiousness, 0.32 for Extraversion, 0.03 for Agreeableness and 0.07 for Neuroticism. Compared to the agreement observed in other domains (e.g., the attribution of emotions), these values are low, but are still statistically significant, i.e. they are not the result of chance. Therefore, the low agreement should not be considered the result of low quality judgments or data, but an effect of the inherent ambiguity of the problem.

9.3.4 APP Experiments: Speech

While being characterized by low agreement levels, social perception experiments still show that the observers tend to associate stimuli and judgments in a coherent way. This can be seen through the performance of APP experiments where the features (physical measurements extracted from the stimuli) are mapped into the judgments of the observers using statistical approaches. The results—the correct prediction rate is above what can be expected by chance—show that the relationship between features and judgments is robust enough to be modeled in statistical terms. In particular, this section focuses on experiments performed over the SSPNet Speaker Personality Corpus (see Sect. 9.3.2) and fully described in [13].

The experiments follow the scheme depicted in the technical layer of Fig. 9.1. The first step is the *feature extraction*, i.e. the transformation of the raw data (speech samples in this case) into vectors of physical measurements expected to account for the main characteristics of speech. Such a format is particularly suitable for computer processing because it allows one to use statistical approaches for the prediction of the traits attributed to the speakers talking in the samples of the corpus.

In the case of the experiments presented in this section, the feature extraction process follows the approach typical of computational paralinguistics [20]. First, short-term features are extracted from 30 ms long audio segments that slide over a speech sample at regular time steps of 10 ms. The short-term features of the experiments are as follows:

- *Pitch*: the fundamental frequency of a human voice and the individual measurement that influences most the way speech sounds;
- *Formants*: the resonances of the vocal trait, an indication of the phonemes being uttered (in the experiments, only the first two formants are used);
- *Energy*: a measure of how loud a speech segment is perceived to be;
- *Speaking rate*: the length distribution of voiced and unvoiced segments-segments where there is or there is not emission of voice, respectively—is used as an indirect measure of how fast a person speaks.

The short-term features are extracted using *Praat*, one of the most popular and commonly applied speech analysis tools [3].

Since short-term features are extracted from short audio segments, the same feature (e.g., the pitch) is measured multiple times for the same sample. Hence, the different values of the same feature are represented in terms of their statistical properties. In the case of the experiments of this section, the statistical properties are as follows:

- *Mean*: average of all the feature values extracted from the sample;
- *Minimum*: minimum of all the feature values extracted from the sample;
- *Maximum*: maximum of all the feature values extracted from the sample;
- *Entropy*: entropy of the distribution of the differences between consecutive measurements of the same feature (see [13] for more details).

Given that the short-term features are six and the statistical properties are four, every speech sample is represented with a 24 dimensional feature vector **x**.

Every sample of the SSPNet Speaker Personality Corpus has been annotated in terms of the personality traits attributed to the speakers by 11 observers (see Sect. 9.3.2). This allows one to split the Corpus into two parts: the first includes the samples that, for a given trait, are below or equal to the median. The second includes all the samples that, for the same trait, are above the median. By splitting the Corpus in this way, it is possible to perform a binary classification task where the goal is to predict whether a speaker has been perceived to be above or below median for a given trait. If C^- is the class of the samples below or equal to the median and C^+ is the class of the samples above the median, then the performance metric is the *accuracy*, i.e. the percentage of times the prediction approach correctly assigns a speech sample to its actual class.

The binary classification is performed using a logistic regression that assigns a feature vector $\mathbf{x}$ to class C with the following probability:

$$p(C|\mathbf{x}) = \frac{\exp\left(\sum_{i=1}^{D} \theta_i x_i - \theta_0\right)}{1 + \exp\left(\sum_{i=1}^{D} \theta_i x_i - \theta_0\right)}, \tag{9.3}$$

where D is the dimension of the feature vector and the θ_js are the model parameters. The main advantage of this approach is that the parameters, resulting from the training of the model, weight the different features and provide indirect information about the features having the highest influence on the classifier outcome. The model has been trained with the k-fold approach ($k = 15$) and the same speaker was never present in both training and test set. In this way, the prediction was speaker independent.

The accuracies are reported in Table 9.3, while the logistic regression coefficients are depicted in Fig. 9.2. For every trait, the plots report the values of the θ_i parameters. Table 9.3 shows that the best performances are achieved for Extraversion and Conscientiousness. This is not surprising because these are the two traits that observers tend to attribute with the highest consensus and coherence: "*[...] there are two dimensions that underlie most judgments of traits, people, groups, and cultures [...] the first makes reference to attributes such as competence, agency, and individualism, and the second to warmth, communality, and collectivism*" [8] (*warmth* and *competence* are alternative names for Extraversion and Conscientiousness, respectively).

The psychology of speech perception finds confirmation in the analysis of the weights as well (see Fig. 9.2). In particular, the high weights for the entropy of pitch, formants and energy confirm, at least partially, that "*Rate and pitch variation were the most influential for competence and benevolence, respectively. For competence, one interaction effect (rate by pitch variation) was significant. For benevolence, two interaction effects were significant (pitch variation by loudness, and pitch variation by rate)*" [17]. This confirms that the machine, at least in the case of speech-based traits attribution, seems to replicate social perception mechanisms in humans.

Table 9.3 The table shows the performances achieved over the Speaker Personality Corpus and the Face Personality Corpus

Trait	Speaker Personality Corpus (%)	Face Personality Corpus (%)
Openness	58.6	66.6
Conscientiousness	72.5	59.9
Extraversion	58.7	66.0
Agreeableness	78.5	59.2
Neuroticism	66.1	67.1

The performances are expressed in terms of accuracy, percentage of times the system assigns a sample to the right class

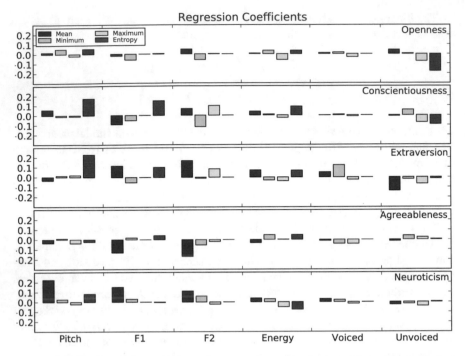

Fig. 9.2 The charts show the coefficients of the logistic regression for the BF Traits. The expressions $F1$ and $F2$ stand for Formant 1 and 2, respectively. For every feature (pitch, $F1$, $F2$, Energy and length of voiced and unvoiced segments), the chart shows the coefficient for four statisticals, namely mean, minimum, maximum and entropy

9.3.5 APP Experiments: Faces

The experiments of this section are similar to those performed on the speech samples of the *SSPNet Speaker Personality Corpus*, but the data is the *SSPNet Face Personality Corpus* (see Sect. 9.3.2) [1]. This includes 829 face images—the frontal pictures of the FERET Corpus originally collected for biometric purposes—annotated by 11 observers that have filled the BFI-10 questionnaire (see Sect. 9.3.1) after watching each of the pictures. The main characteristics of the corpus are available in Table 9.2 and the experiments presented in this section have been described in full detail in [1].

Figure 9.3 shows the pictures that have attracted the lowest and highest scores for each of the BF traits. The difference between the faces at the extremes of each trait suggests how the observers tend to map differences in facial appearance into differences in personality traits. However, only prediction experiments like those of the previous section can say whether the mapping is coherent and, if yes, to what extent.

Following the technical layer of Fig. 9.1, the first step of the process is the feature extraction. In the experiments of this work, the images are first normalized to have all the same size (65 × 50 pixels). Then, the colours are mapped into intensity values

Fig. 9.3 The picture shows, for each trait, the faces that have attracted the lowest (*lower pictures*) and highest (*upper pictures*) rating

by taking the average of the intensities of the three components of each pixel (*Hue, Saturation* and *Value*). The intensity values obtained with such a process are thus used as features. However, given that the dimensionality of the feature vectors is too high ($65 \times 50 = 3250$), the images are projected onto the first 103 Principal Components [2], corresponding to 90 % of the variance in the data.

Like in the case of the speech experiments described in the previous section, the faces of the corpus are split into two classes for each trait, namelys samples above and samples below or equal median according to the scores assigned by the observers. The feature extraction process consists in extracting the projection of the faces onto the Principal Components extracted from the faces of the training set (see [1] for the details). The prediction of the attributed traits has been performed with Support Vector Machines.

Table 9.3 shows the performances achieved for the different traits. In this case, the best performances are achieved for Extraversion, Openness and Neuroticism. The first trait tends to be the best predicted in most of the works presented in the literature [24] because it is one of the two traits that people tend to infer more easily when they are contact with others [8]. For the other two, the performance changes significantly from case to case. This should not be considered a problem of the approach, but a property of the data that make a given trait more or less *available* [27] (see Sect. 9.4 for more details). In any case, the results show that the observers tend to map facial characteristics into traits with sufficient consistency and coherence to allow statistical modeling.

9.4 Discussion and Conclusions

This chapter has addressed the problem of social perception in machines. In particular, the chapter has shown two examples of Automatic Personality Perception, i.e. the task of predicting the traits that observers attribute to subjects they enter in contact with. Besides showing that machines can replicate the way human observers perceive the personality of others, the examples illustrate two major peculiarities of social perception experiments: The first is that the goal of the experiments is not to predict the *"true"* traits of the targets, but the traits that other attribute to them. For this reason, the experiments are often characterized by a low, but statistically significant degree of agreement between observers. The second is that not all traits can be predicted equally well from the same data, the reason being that different data make different traits more or less *available*, i.e. more or less easy to infer for human observers.

Low agreement between observers is typically seen as a problem, but in the case of social perception it should be considered a consequence of the inherent ambiguity of the problem. The Brunswik's Lens (see Fig. 9.1) shows that actual and attributed states are not necessarily the same. In this respect, the *accuracy* of attribution processes should not be considered a goal to be achieved, but simply a characteristic of a given type of perception. When the externalization processes allow the observers to easily infer the actual states of the targets, then the observers tend to be more accurate and, as a consequence, to reach higher levels of agreement. When the externalization processes do not allow the observers to infer the actual state, then the accuracy is lower and, as a consequence, the agreement is lower as well. In the case of the Big-Five, perception studies show that the median of the agreements observed in different studies range roughly between 0.1 and 0.3 [10, 11] and they are accepted as long as they are statistically significant, i.e. they are not the result of chance.

Differences in performance depending on the data are another characteristic of social perception experiments. In the particular case of the BF, this means that different traits are predicted with different performances in different corpora. In this case as well, the phenomenon should not be seen as a problem, but an effect of the availability [27], i.e. of how much the data make a trait (or any other state) accessible to inference. The two examples presented in this chapter show that Extraversion can be predicted effectively with both speech and faces—not surprisingly given that this trait is probably the most easily available [8, 24]—while the others are more or less accessible depending on the corpus. Very likely, these differences can provide useful information about the channels through which people actually develop their impressions and social perception inferences.

Social inferences play a major role in everyday interactions. Impressions about others influence, to a significant extent, the way people behave towards others [22]. Furthermore, the social identity of individuals depends not only on their actual characteristics, but also on the characteristics attributed by others [7]. For these reasons, automatic approaches for social perception can help to develop socially believable behaving agents and, not surprisingly, automatic social perception is the subject of major efforts in the computing community.

References

1. Al Moubayed N, Vazquez-Alvarez Y, McKay A, Vinciarelli A (2014) Face-based automatic personality perception. In: Proceedings of the ACM international conference on multimedia, pp 1153–1156
2. Bishop C (2006) Pattern recognition and machine learning. Springer, New York
3. Boersma P (2002) Praat, a system for doing phonetics by computer. Glot Int 5(9/10):341–345
4. Brunswik E (1956) Perception and the representative design of psychological experiments. University of California Press, Berkeley
5. Funder D (2001) Personality. Annu Rev Psychol 52:197–221
6. Howell D (2012) Statistical methods for psychology. Cengage Learning, Belmont
7. Jenkins R (2014) Social identity. Routledge, New York
8. Judd C, James-Hawkins L, Yzerbyt V, Kashima Y (2005) Fundamental dimensions of social judgment: unrdestanding the relations bteween judgments of competence and warmth. J Pers Soc Psychol 89(6):899–913
9. Kang S, Bodenhausen G (2015) Multiple identities in social perception and interaction: challenges and opportunities. Annu Rev Psychol 66:547–574
10. Kenny D (2004) PERSON: a general model of interpersonal perception. Pers Soc Psychol Rev 8(3):265–280
11. Kenny D, Albright L, Malloy T, Kashy D (1994) Consensus in interpersonal perception: acquaintance and the big five. Psychol Bull 116(2):245–258
12. Kunda Z (1999) Social cognition: making sense of people. MIT press, Cambridge
13. Mohammadi G, Vinciarelli A (2012) Automatic personality perception: prediction of trait attribution based on prosodic features. IEEE Trans Affect Comput 3(3):273–284
14. Nass C, Brave S (2005) Wired for speech: how voice activates and advances the human-computer relationship. MIT press, Cambridge
15. Pantic M, Cowic R, D'Errico F, Heylen D, Mehu M, Pelachaud C, Poggi I, Schroeder M, Vinciarelli A (2011) Social signal processing: the research agenda. In: Visual analysis of humans. Springer, London, pp 511–538
16. Rammstedt B, John O (2007) Measuring personality in one minute or less: a 10-item short version of the big five inventory in English and German. J Res Pers 41(1):203–212
17. Ray GB (1986) Vocally cued personality prototypes: an implicit personality theory approach. J Commun Monogr 53(3):266–276
18. Rosenthal R (2005) Conducting judgment studies: some methodological issues. In: Harrigan J, Rosenthal R, Scherer K (eds) The new handbook of methods in nonverbal behavior research. Oxford University Press, New York, pp 199–234
19. Saucier G, Goldberg L (1996) The language of personality: lexical perspectives on the five-factor model. In: Wiggins J (ed) The five-factor model of personality. Guilford Press, New York, pp 21–50
20. Schuller B, Batliner A (2013) Computational paralinguistics: emotion, affect and personality in speech and language processing. Wiley, Hoboken
21. Todorov A, Mandisodza A, Goren A, Hall C (2005) Inferences of competence from faces predict election outcomes. Science 308(5728):1623–1626
22. Uleman J, Adil Saribay S, Gonzalez C (2008) Spontaneous inferences, implicit impressions, and implicit theories. Annu Rev Psychol 59:329–360
23. Uleman JS, Newman LS, Moskowitz GB (1996) People as flexible interpreters: evidence and issues from spontaneous trait inference. In: Zanna MP (ed) Advances in experimental social psychology, vol 28, pp 211–279

24. Vinciarelli A, Mohammadi G (2014) A survey of personality computing. IEEE Trans Affect Comput 5(3):273–291
25. Vinciarelli A, Pentland A (2015) New social signals in a new interaction world: the next frontier for social signal processing. IEEE Syst Man Cybern Mag (to appear)
26. Vinciarelli A, Esposito A, André E, Bonin F, Chetouani M, Cohn J, Cristani M, Fuhrmann F, Gilmartin E, Hammal Z, Heylen D, Kaiser R, Koutsombogera M, Potamianos A, Renals S, Riccardi G, Salah A (2015) Open challenges in modelling, analysis and synthesis of human behaviour in human-human and human-machine interactions. Cogn Comput (to appear)
27. Wright A (2014) Current directions in personality science and the potential for advances through computing. IEEE Trans Affect Comput (to appear)

Chapter 10
Towards Modelling Multimodal and Multiparty Interaction in Educational Settings

Maria Koutsombogera, Miltos Deligiannis, Maria Giagkou and Harris Papageorgiou

Abstract This paper presents an experimental design and setup that explores the interaction between two children and their tutor during a question–answer session of a reading comprehension task. The multimodal aspects of the interactions are analysed in terms of preferred signals and strategies that speakers employ to carry out successful multi-party conversations. This analysis will form the basis for the development of behavioral models accounting for the specific context. We envisage the integration of such models into intelligent, context-aware systems, i.e. an embodied dialogue system that has the role of a tutor and is able to carry out a discussion in a multiparty setting by exploring the multimodal signals of the children. This system will have the ability to discuss a text and address questions to the children, encouraging collaboration and equal participation in the discussion and assessing the answers that the children give. The paper focuses on the design of the appropriate setup, the data collection and the analysis of the multimodal signals that are important for the realization of such a system.

Keywords Multiparty and multimodal interaction · Reading comprehension · Non-verbal signals · Turn taking · Feedback · Embodied dialogue system

M. Koutsombogera (✉) · M. Deligiannis · M. Giagkou · H. Papageorgiou
Institute for Language and Speech Processing - "Athena" R.I.C, Athens, Greece
e-mail: mkouts@ilsp.athena-innovation.gr

M. Deligiannis
e-mail: mdel@ilsp.athena-innovation.gr

M. Giagkou
e-mail: mgiagkou@ilsp.athena-innovation.gr

H. Papageorgiou
e-mail: xaris@ilsp.athena-innovation.gr

© Springer International Publishing Switzerland 2016
A. Esposito and L.C. Jain (eds.), *Toward Robotic Socially Believable Behaving Systems - Volume II*, Intelligent Systems Reference Library 106,
DOI 10.1007/978-3-319-31053-4_10

10.1 Introduction

The study described in this paper explores interaction in an educational context, where a tutor holds a conversation with two children, asking them questions on texts they have studied and assessing their reading comprehension through these questions. Furthermore, the tutor encourages collaboration between the children and helps them to identify the correct answers within the text. The main goal of this investigation is to model the conversational behavior of participants in such a context, in view of a virtual agent that supports reading in educational settings and fosters comprehension and retention of texts by young learners through interaction.

Written texts are one of the commonest means of delivering information in educational settings, hence effective reading instruction is of utmost importance for knowledge acquisition. At the same time, the fact that most reading takes place on electronic devices (pc, tablets, mobile phones) either online or offline, poses another challenge for the study at hand, i.e. the investigation of reading support and tutoring on electronic devices.

In our view, this challenge could be addressed in a smart and adaptive learning environment in the core of which lies a dialogue system embodied by a virtual agent. The agent interacts with a pupil, assists her/him in studying a factual text and is able to assess reading comprehension, guiding the learner through the text, asking questions and helping the learner identify and clarify information-dense excerpts. Furthermore, we envisage an agent that is able to interact with more than one child at a time, i.e. at least a pair of children; in this respect, this agent should be able to exhibit social skills, but, also to monitor the social skills of its co-locutors.

A virtual agent that interacts with learners is a human-like interface that should be able to understand and communicate the verbal and non-verbal actions of the participants, i.e. the multimodal signals that provide dynamically significant information about their state of mind [1–3]. Learning is a scenario in which conversational, social and affective participants' skills influence the outcome of the interaction and its success [4–6]. Consequently, one of the key aspects to be incorporated in the system is the modelling of the interaction between the participants; an aspect that becomes more challenging as the number of speakers in an interaction grows and the communication dynamics become more complex.

The design and implementation of such an intelligent system is heavily based on the modelling of the communication among the participants in this specific context. Thus, this work attempts to provide preliminary insights of how a model of interactional behavior should be formed, in order to feed and equip an intelligent interface that can be used in educational tasks, such as in individual or group tuition, or in homework assistance.

The study of human–human interactions sheds light on actions and strategies participants employ to convey their message and achieve their communicative goals, as well as the modalities involved. Recently, the availability of robust capture devices (such as eye trackers, microphones, biosignal sensors) and modelling techniques has been exploited to build dialogue systems embodied by virtual agents or robots

[7–9]. A further challenge that is addressed is to approach multiparty and multi-modal conversations, which are a more complex construct compared to two-party conversations, in terms of turn management, feedback responses and variables such as engagement, attention and dominance. These are crucial features to the interaction strategies and the regulatory actions of an agent within a dialogue system for educational purposes.

The setup and task presented here aim at compiling a corpus of tutor–students interactions which will facilitate the investigation of the conversation structure, the coordination of turn-taking and the multimodal feedback and attention signals the speakers employ. The annotations on the data aim at modelling the multimodal and multiparty interaction towards the implementation of an embodied dialogue system able to handle more than one co-locutor, discuss with them and encourage their collaboration, engage them in the task and assess their reading comprehension.

The experimental setup and data analysis presented in this paper aim to define and discuss (a) the appropriate scenario and task design, i.e. the material and the spatial set-up enabling to capture the desired behavioural cues, (b) aspects of multimodal interactional behavior that denote the speakers' reactions and their conversational strategies, and (c) association of the reading comprehension task to the reader's verbal and non-verbal activity.

The remainder of the paper discusses the experimental design and set up that led to the creation of the data collection. Next, we describe the annotation process performed on the collected corpus to reflect the aspects and features of interest. The analysis of the annotated sessions is presented in Sect. 10.4. Finally, the discussion section includes aspects of multimodal and multiparty interaction behavior and flow that are highlighted through the analysis and are important to be modelled in an HCI dialogue interface, with promising results for this challenging area.

10.2   Experimental Design and Data Acquisition

To be able to make observations of human–human interactions, we followed the best practice, i.e. aimed to the acquisition of audiovisual recordings. Large-scale multimodal corpora have been developed to advance the understanding of human–human interaction [10–13] and model verbal and visual signals in interaction to build more engaging human–machine interfaces, including social, natural and intelligent dialogue systems. Such resources are a prerequisite to the behavior and interaction modelling.

Our data collection consists of four sessions of overall 41 min duration. In each session, an adult tutor coordinates the discussion with two children on a text that they have already read right before each session started. The tutor poses a set of questions on the text and the children have to answer them either directly or after collaborating with each other and reaching an agreement. The number of participants and the spatial arrangements were designed in a way to better serve the purposes of the analysis, but also to create a semi-controlled set-up that can be replicated and modelled at a later stage. The choice of more than one child was done to study the way multi-party interaction is performed; however, the presence of more than two children would significantly increase the complexity of the task analysis, since each signal would have to be interpreted by checking actions or reactions towards several directions, i.e. as many as the number of participants. This set-up is not intended to be a simulation of the classroom setting, but of other teaching environments, such as group tuition or homework assistance.

The spatial set-up was implemented with the three participants sitting around a table, holding an equal distance between them so that their signals would be visible when addressing another participant.

In designing the task, selecting the texts, and forming the core comprehension questions to trigger interaction, a set of parameters and constraints were taken into account: (a) the learners' profile, in terms of age, language development and cognitive maturity, (b) the specificity of the topic which should provide the grounds for an interaction manageable by a dialogue system, i.e. question–answer pairs that are factual and narrow enough to allow objective assessment, and (c) the encouragement of collaboration between participants. Under these constraints, the material used for the experiments were two texts on the subject of geography. The texts were drawn from textbooks addressed to the specified age groups (first and second grades of the secondary school) in order to ensure level-appropriateness. Non-fiction informative texts were selected because of their factoid nature which allows for the objective automated assessment of comprehension. More specifically, geography texts usually employ a set of associated named entities linked with relations of part-whole, cause-effect, order/sequence/direction etc., thus allowing for the assessment of both low- and high-level cognitive processes involved in comprehension through the following types of questions: yes–no questions, wh-questions, check and list questions.

The recordings were carried out at the ILSP/Athena recording studio, during the pre-event of Researcher's Night.[1] A call for participation under the title "Let's design the future tutor"[2] was posted on the event webpage and addressed children of the first and second grades of secondary school. Eight children were recruited for whom signed parental consent was obtained. The participants were put in pairs according to their school grade. Balanced groupings of participants in terms of gender, personality and performance would have been desirable; however at this stage, grouping was random and based on the subjects availability during the event.

The person that undertook the role of the tutor in all sessions is a speech therapist well-acquainted with interacting with children. We conventionally refer to her as "tutor", since the focus is on her conversational role in the particular set

---

[1] https://www.athena-innovation.gr/en/announce/pressreleases/232-ren2014.html.

[2] "What do you consider an ideal tutor? How would you like a tutor-avatar or a tutor-robot? We'll guide you through our recording studio where we'll record the way a tutor communicates with the students with the goal of designing a virtual tutor. Can you help us design the tutor of the future? All you have to do is to answer a few simple questions and you'll have the chance to see how we collect the data that are necessary to develop a virtual tutor."

of interactions, and was perceived as such by the children. Four groups were formed and four sessions were recorded during the event. What follows next is a typical structure of the conversation carried out in each session.

10.2.1   Session Design

Preparation phase: Each pair read one of the following texts: (i) Text 1 "The earth's biggest mountain ranges" (595 words), followed by 9 questions. (ii) Text 2 "The Baltic and the North Sea" (515 words), followed by 6 questions. Different types of questions were designed to attest whether they elicit different kind of answers and consequently whether a dialogue system would be able to handle and assess all the kinds of answers given.

The tutor was contacted in advance and was given both texts and their questions as well as instructions on how to coordinate the interaction with each group of children.

Before each session, the children were given 10 min to go through their text and read it as many times as they wanted to, without discussing it with their partner.

Introduction to the activity: The tutor introduced herself and asked the learners what their names were. The tutor welcomed and explained the aim of the session. A brief phase of small talk then followed aiming at making the children feel at ease (e.g. the school the children went to, whether they knew each other etc.). The tutor then described the activity and the subsequent steps, i.e. that she was about to ask simple questions on the text and that the children may reply on their own, or after discussing and collaborating with each other to conclude on a specific answer.

Question and answer phase: The tutor posed the questions on the text one at a time in a predefined order and elicited the answers from the children. The following cases were specified:

1. As soon as a student replied, the tutor's role was to ask the other one whether he/she agreed with it and tried to initiate a conversation. If there was disagreement, then the tutor asked the children to discuss their responses with each other and conclude with a final answer.
2. If the reply was correct, and an agreement had been reached, the tutor verbally congratulated the student(s) and moved on to the next question.
3. If the reply was not correct, or if there was a delay in the answer, or there was disagreement between the 2 students, or following a discussion between the pupils they admitted that they did not know the answer, the tutor read the respective excerpt from the text which contained the correct answer. The tutor asked the pupils the question again, the goal here being to elicit the answer from the pupils (and not provide the answer herself) to make sure that they had understood the text. The tutor was given further general instructions on her role and how to dynamically manage the flow of the interaction and the turn-taking according to the children's conversational behavior, ensuring at the same time equal participation. Specifically:

- The tutor was instructed to make sure that she did not look at the text, at least not all the time. She also had to avoid gesturing or touching her face, so that all facial characteristics were visible to be annotated during the analysis.
- The first 2 questions were not supposed to be addressed to a particular child. The tutor was supposed to give enough time to the children to respond as they wished, or to discuss the answer.
- If during the first 2 questions one of the 2 students was very talkative and monopolized the discussion, the tutor was instructed to address the next question to the other child.
- If one of the 2 students kept being more extrovert or conversationally dominant, then the tutor was supposed to address one of the forthcoming questions to the less talkative child.
- When pupils were discussing a question, the tutor was not supposed to interfere in the discussion within a certain time threshold; she was rather supposed to wait till the discussion was completed and the pupils addressed her again.

Closing phase. After all the questions had been discussed, the tutor thanked and greeted the students.

The sessions recorded include the introduction to the activity, the question–answer phase and the closing phase. These sessions constitute the raw data of the multimodal corpus (Table 10.1)

Technical setup. All sessions were recorded using three high definition cameras and one directional microphone. Each camera was used to capture a different angle, i.e. the frontal view of each speaker. The camera directed at the tutor had a wide frame to cover the entire scene, and a zoomed frame was used to study the tutor's behavior in detail.

Physical setup. The three participants were sitting around a round table and there was an equal distance between them (cf. Fig. 10.1) so that their head turns when addressing another participant were visible.

Table 10.1 Corpus description

Session	Session duration	Text	Children gender (male/female)
1	7.20	Text2	M–M
2	11.43	Text1	M–M
3	12.10	Text2	F–F
4	9.50	Text1	M–F

Fig. 10.1   A photo of the experimental setup with all 3 participants

10.3   Annotation Process and Scheme

The recorded sessions were annotated by an expert annotator and were partly val-
idated by another expert. The focus was set on describing the form and functions
of the verbal and non-verbal signals employed in the multimodal strategies of the
conversation participants. The sessions were annotated using the ELAN[3] editor [14]
and several tiers were introduced to cover all annotation aspects. For each session,
all 4 videos covering different angles and the audio were synchronized so that the
annotator had a clear view of not only each participant's behavior, but of the whole
interaction and the interplay between participants as well.

The annotation scheme that was employed contained features and values that
needed to be represented for the analysis of the conversational behavior of the participants. It is based on the MUMIN scheme for the encoding of multimodal interaction
[15], which had been tailored for the needs of the specific task and has been used in
similar studies of multiparty communication [16]. The goal of the annotation was to
account for conversational multimodal behavior through verbal and nonverbal signals that have specific forms and communicative functions. Thus, a large part of the
scheme is dedicated to the annotation of the non-verbal modalities of the participants. In this study, the focus was on head movements, facial expressions and facial
gestures, cues that are considered important to the regulation of the interaction and
feedback expression. Each of these signals was first identified on the time axis, its
form was marked and it was assigned a feedback and a turn management value. The
annotation layers included in the scheme were as follows:

---

[3]ELAN (http://www.lat-mpi.eu/tools/elan/).

Structure and sub-structure. The conversation was split into structural units where each structure item reflected the kind of the conversational action and its function, i.e. introduction, questionX, answerX, conclusion. This kind of segmentation and labeling was used instead of dialogue act labels in order to be closer to a set of events that a dialogue system would use. The identification of question and answer sections is important for the modelling as it includes utterances that are specific to the subtopic discussed. An introduction and conclusion were used for the opening and closing of the session, respectively. QuestionX was assigned to segments where a question was addressed, along with potential prefatory statements, while answerX corresponded to segments where the answer or the discussion that lead to an answer took place.

Speech activity. Under each structure segment, the utterances of all participants were transcribed and synchronized with the video stream. The speakers' transcriptions were collected to attest different patterns and alternatives one may use to instantiate the particular set of structural segments, i.e. questions and answers, introduction and closing and how these patterns may be exploited to distinguish between different structural segments. This was more straightforward for the case of the tutor, who, due to her coordinating role, expressed a more or less pre-defined content, without however being committed to use a specific sequence of words and sentences.

Turn management. Turn-taking is a substantial mechanism for the organization of turns in conversation [17], of turn regulation (i.e. head gestures, gaze) [18], as it correlates with attention [19] and is a prerequisite in multiparty dialogue systems in order to handle floor coordination [20]. The mechanism accounts for forms of turn allocation, rules that apply in transition-relevant places as well as aspects of collaborative and non-collaborative interactions such as interruptions and overlap resolution devices, both from verbal and nonverbal perspectives.

From the annotation point of view, this layer contains values that describe the way speakers regulate the interaction and the succession of turns by verbal and non-verbal means. Specifically, there are values for taking (*take, accept, grab*), keeping (*hold*), or releasing (*offer, complete, yield*) the turn. The distinct value of backchannel is also included to differentiate backchannel cues from meaningful signals. Turn management is considered here a conversational function applicable to all signals that speakers employ. Hence, each layer of verbal and non-verbal activity is accompanied by a turn management relevant function.

Feedback. Feedback responses are realized through verbal and non-verbal means to show perception and understanding of a message, agreement or disagreement and they support the collaborative aspects of dialogue in terms of acknowledgement and continuation. Feedback has been viewed through a robust theoretical framework [21] and from a multimodal point of view with respect to cues inducing listeners' feedback, among others [22]. Moreover it has been described as occurring in different social activities or modelled for the purpose of behavior simulation [23].

Same as above, values related to feedback are attributed horizontally to cover both functions of verbal and non-verbal feedback instantiations, i.e. either via back-channels and expressing evaluations, or through non-verbal actions such as nodding, smiling or other head movements and facial expressions. The related labels describe the kind of feedback the speakers give or elicit in terms of perception, understanding, acceptance or not. Feedback signals are important for the continuation of the discussion and the acknowledgment of each speaker's conversational actions and content.

General facial expression. The speakers' generic facial expressions of smile, laugh and scowl indicate their perception of the discussion and their state of mind towards its content. They are employed to show agreement, encouragement and satisfaction, or, on the contrary, doubt, disagreement or unpleasantness. Subsequently, they trigger the co-locutors' responses accordingly.

Head movement. Speakers move their heads in forms and directions that are significant to the establishment of feedback and turn regulation functions. Head *nodding*, for example, is employed to acknowledge perception agreement with what is being said; head *turn* may determine the attention focus of the speakers since it is linked to their gaze direction; head shake may be a sign of disagreement or doubt, while *tilting* the head or moving it *forward* and *backward* may be employed to reinforce a speaker's message.

Gaze. Studying gaze in multi-party interaction enables the identification of the addressee of an utterance, the focus of attention, or the next speaker allocation. Hence, gaze direction becomes an important and helpful signal from which one may infer the turn coordination as well as the target of attention. The annotation scheme contains values of attentive gaze towards the tutor and either of the two children. These values are attributed whenever a communicative, according to the criteria just mentioned, gaze shift occurs, independently of whether a speaker has the floor or not: e.g. a participant may shift the gaze while talking to make clear who the addressee is or to elicit feedback; gaze shifts of a listener may be employed to offer or elicit feedback. A value is also included for cases where participants gaze away, i.e. to show they are thinking.

Eyes. Eye openness may signal variations of emotion or engagement, e.g. surprise or enthusiasm (*wide open*), as well as contemplation, interest, attention or disagreement (*semi-closed, blink*).

Eyebrows. Moving eyebrows may indicate involvement, encouragement, attention and surprise (*raise*) or doubt, disagreement or contemplation (*frown*) of the participants.

Mouth. Participants with *closed* mouth with protruded lips may be expressing agreement feedback signals, especially if they are coupled with head nodding, while an *open* mouth is annotated as a sign that a participant is attempting to take the turn when another speaker has the floor.

The annotation scheme is described in Table 10.2 while Fig. 10.2 shows a snapshot of the ELAN annotation editor including all 4 views of the setting that the annotator has at his disposal.

Table 10.2 The coding scheme used for the annotation of the 4 sessions

Annotation layers	Values
Speech	Speech transcription
Structure	Introduction, sections, closing
Substructure	Introduction, closing, questionX, answerX
Turn management	Take, accept, grab, offer, complete, yield, hold, backchannel
Feedback	Perception/understanding (give-elicit) Accept (give-elicit), Non-accept (give-elicit)
Face_general	Smile, laugh, scowl
Functions_Face	Feedback, turn management
Head_movement	Nod(s), shake, jerk, tilt, turn, forward, backward
Functions_Head_movement	Feedback, turn management
Gaze	Attention_Child_A, Attention_Child_B, Attention_Tutor, Gaze_away
Functions_Gaze	Feedback, turn management
Eyes	Wide_open, Semi-closed, one-closed, blink, closed
Functions_Eyes	Feedback, turn management
Eyebrows	Raise, frown
Functions_Eyebrows	Feedback, turn management
Mouth	Open, closed
Functions_ Mouth	Feedback, turn management

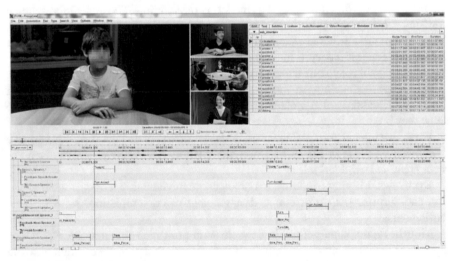

Fig. 10.2 A screenshot of the ELAN annotation editor with the videos and the synchronized manual annotations. All 4 angles of the same time frame may be available to the annotator

10.4   Data Analysis

The annotation process of the recorded sessions resulted in a rich number of annotations that were analysed to discover (a) forms and modalities that the participants
exploited during interaction, (b) frequent or preferred strategies in conversation management (c) feedback mechanisms that speakers employed to show a variety of reactions, and (d) patterns of behavior that may occur in a specific context, such as in a
question–answer sequence of educational content between a tutor and two children.
Special attention was also paid to the parameter of time, i.e. either the duration of
a certain behavior, or the succession of specific events in the course of discussion
and the timed responses of the participants. At this status of the study, the focus is
on low-level signals (i.e. verbal and non-verbal modalities) as well as some preliminary attempts to correlate different signals and find possible dependencies. Further
investigation is expected to shed light on the association between specific signals
and the subsequent responses, as well as on inferences about high-level signals of
interactional behavior, such as engagement and dominance.

10.4.1   Speech Duration

Speech duration was calculated for each participant to determine the distribution of
their speech activity among speakers. Unequal speaking times among children may
be a sign of dominance. To get more information about speech activity, speaking
times were also compared to the number of utterances produced by each participant
to understand whether a speaker was talkative or not. A balanced number of utterances
among 2 speakers shows that they may equally take the floor, even though the absolute
speaking times of one speaker are higher than the other, which might occur because
one of the 2 speakers makes shorter contributions. On the other hand, unbalanced
numbers of utterances and unequal speaking times may be due to the content, i.e. a
specific answer to a question needs elaboration and a lot of wording. This information
is also important since it describes the extent of a child's contribution to a given
question.

   Figure 10.3 shows the percentage of the speech duration for each speaker role, i.e.
children and tutor, and the average number of utterances each speaker role produces
per second. In all cases, the tutor had the highest speech activity, which seems to
be explained by the role she assumed as the coordinator of the interaction and the
participant who addressed both speakers, helped, encouraged, checked the answers,
etc. The tutor also had the lowest frequency of utterances per second, which makes
sense given that she often produced long and explanatory utterances.

   On the part of the children, the relation of speaking times to number of utterances
has been consistent throughout the sessions, i.e. the less the speech activity, the
higher the number of utterances per second. This means that children who were less
talkative either made shorter contributions or they spoke to the point. However, to

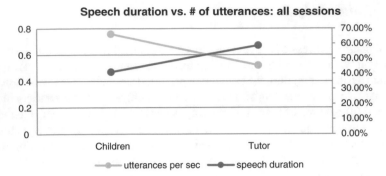

Speech duration vs. # of utterances: all sessions

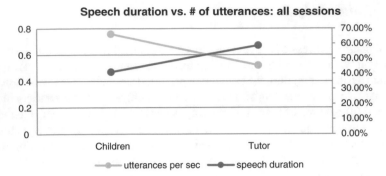

Fig. 10.3 The relation between percentage means of speech activity per speaker role (Children, Tutor) and the average number of utterances/s. speakers produced in all sessions

be able to reach firm conclusions, more examples of interactions are needed to study the relation between speech activity and the number of utterances produced.

10.4.2 Turn Management

The analysis of turn management annotation indicates the modalities that speakers employed and the respective turn actions that were performed. The volume of turn management labels in our data amounts to 1350; 64 % of them was attributed to the tutor and 36 % to the children. All sessions presented a normal flow in the transition of turns, resulting in a high number of values such as turn offer and turn accept (cf. Fig. 10.4), which are representative of prototypical turn-taking. Apart from speech, the preferred non-verbal modalities for turn regulation were head movements and gaze (cf. Fig. 10.5). According to the structure of the discussion, the most frequent pattern was that of the tutor offering a turn and a child accepting it (cf. Fig. 10.6).

Fig. 10.4 Distribution of turn management values in all 4 sessions

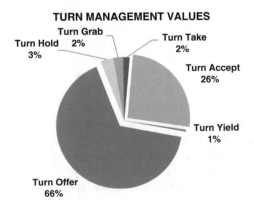

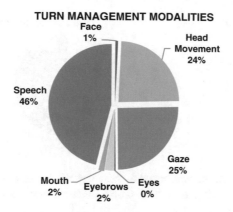

Fig. 10.5 Distribution of turn management modalities in all 4 sessions

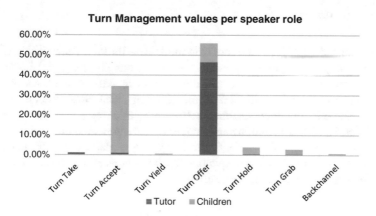

Fig. 10.6 Distribution of turn management values per speaker role in all 4 sessions

The interaction was framed in an educational context, meaning that the tutor controlled to a certain degree the conversation by posing questions and enabling both children to collaborate. It is worth noting that due to the collaborative nature of the dialogue, there are few overlaps. Specifically, out of the 25 segments (20 attributed to children, 5 to tutor) that were annotated with the turn grab value, only 12 % of them refer to pure interruptions, while the rest of the percentage is about simultaneous speech due to collaborative completions, usually exploiting a delay or a pause of a participant.

What is challenging in this setting, though, is the multiparty dimension regarding the way turn allocation was performed by the tutor and the respective responses from the children. Given the central role of the tutor, we examined the forms in which she performed turn allocation (cf. Fig. 10.7). The majority of turn offers was performed using both speech and non-verbal modalities. However, it is interesting

Fig. 10.7 Distribution of
forms of turn allocation
performed by the tutor

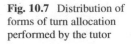

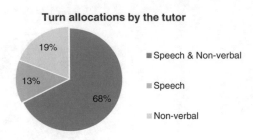

that 19 % of turn allocation was based on non-verbal signals only, denoting that non-verbal modalities in multiparty conversations have an important role within the turn management mechanism.

10.4.3 Feedback

Annotated instances of feedback show the modalities through which it is performed and what it expresses through specific values. In our dataset, 3090 instances of feedback were annotated that correspond to a set of modalities and functions (48 % was attributed to the tutor and 52 % to the children). The most frequent functions are giving and eliciting understanding, followed by eliciting and giving acceptance feedback (cf. Fig. 10.8). Children most frequently gave understanding, while the tutor elicited understanding (cf. Fig. 10.9). Similarly, the modalities usually involved in feedback are, from most to less frequent, gaze, head movements, facial expressions, mouth and eye movements (cf. Fig. 10.10).

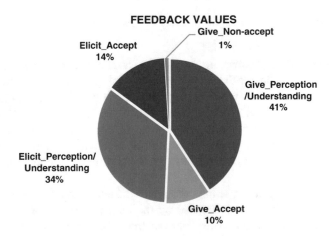

Fig. 10.8 Distribution of feedback values in all 4 sessions

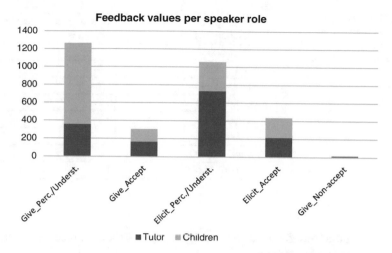

Fig. 10.9 Distribution of feedback values per speaker role in all 4 sessions

Fig. 10.10 Distribution of feedback modalities in all 4 sessions

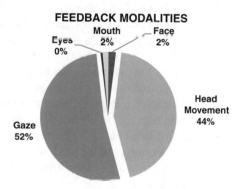

Head movements, especially nods, were primarily used for back-channeling and acknowledgements and were particularly employed by the tutor, who was supposed to be responsive to the children. Such modalities were usually coupled with the offer or elicitation or feedback. Gaze shifts were also the most frequent mechanism to provide signals of perception not only of the message, but of the interaction flow as well.

Feedback signals were important to the dynamics of the conversation. Signs of non-understanding or disagreement on the part of the children usually prompted an action from the tutor, who provided clarifications or allocated the turn to another speaker. Signals of acknowledgement and agreement on the part of the tutor provided encouragement and continuation of the discussion.

The frequency of feedback occurrence was also another important factor for modelling the tutor's behavior, i.e. the frequency with which a virtual tutor should generate silent feedback responses to look more natural. The tutor's feedback frequency was co-examined with the duration of speech activity of the children. In Table 10.3 below

	Child speech activity duration (in ms)	Tutor avg. no. of feedback responses
Table 10.3 Feedback responses from tutor to children according to their speech activity duration	Up to 500	1
	500–1.000	1.4
	1000–1500	1.8
	1.500–2.000	2
	2.000–3.000	2.5
	3.000–5.000	3
	5.000–10.000	4.5
	10.000–20.000	7

we list a set of time thresholds and the average amount of the tutor feedback responses that were observed in our data.

Feedback is a reaction to a conversational action or verbal content that another speaker performs. To better represent the data and enable feedback modelling, we distinguish between feedback source (participant that performs a feedback action), target (participant to whom feedback is given or from whom feedback is elicited) and trigger, i.e. the participant that causes feedback to occur, who, in multiparty interaction may be other than the target, a third party who is also a listener at that moment. For example, in a case where the tutor has the floor and child A turns to child B to give or elicit feedback, the tutor is the trigger, child A is the source and child B is the target. In our data we observed that the percentage of feedback responses where the trigger is different from the target accounted for 13 % of all feedback instances. In this respect, the target of feedback seems to be an important factor when modelling conversational dynamics, since it might influence the sequence of the conversational actions that follow.

10.4.4 Gaze Analysis

One of the most significant non-verbal signals of the interaction flow is that of gaze. It shows the focus of attention of the conversation participants and it is very often employed to signal turn allocation. Moreover, it has a strong feedback function, providing acknowledgments to a speaker and signs of perception and continuation.

At the same time it may function as a dominance mitigation mechanism: a frequent case observed in the dataset is that of child A holding the floor more often than child B; the gaze shifts of the tutor from child A to child B imply that child B should contribute more, and it is an implicit turn offer to child B. We examined these cases by exploiting the feedback trigger-target distinction, in which tutor gazes had a feedback function and the trigger (e.g. child A) was different from the target (child B). About 80 instantiations of this behavior were found and 14 % of them accounted for behaviors

Table 10.4 Relation between the percentage of utterances dominant (DC) and non-dominant (N-DC) children produced and the number of tutor gazes (T-gaze) addressing the respective speaker type

	DC (%)	N-DC (%)	Diff (%)	T-gaze DC (%)	T-gaze N-DC (%)	Diff (%)
Session 1	56.9	43.1	13.7	50.9	49.1	1.8
Session 2	56.8	43.2	13.7	50.4	49.6	0.8
Session 3	51.9	48.1	3.9	52.0	48.0	4.1
Session 4	63.7	36.3	27.3	50.6	49.4	1.2

where the target perceived the feedback signal and eventually took the floor that was implicitly offered by the tutor.

We also calculated the relation between the percentage of utterances that a child produced and the amount of tutor gazes targeting the respective child (cf. Table 10.4). The question here is whether the tutor gazes are consistent with the speech activity of the children or, on the other hand, whether gazes are a mechanism for encouraging less talkative children to participate more or to engage them in the conversation by silently eliciting feedback through gaze. For this purpose, we considered the children that produced more utterances in each session as dominant (DC) and the other half of each pair as non-dominant (N-DC) speakers. Subsequently, we calculated the percentage of the tutor gazes (T-gaze) towards dominant and non-dominant children respectively. Eventually, judging by the difference in tutor gazes towards the two types of speakers, there are no consistent findings coming from the data, i.e. there is no solid evidence regarding this kind of correlation. Again, we believe that more data, i.e. more examples of this behavior would allow for certain conclusions of whether (a) a tutor addresses the gaze equally to both participants independently of their speech activity, implying through gaze that he fosters equal participation, or (b) the number of tutor gazes grows along with the amount of utterances produced by the children.

10.5 Discussion and Future Work

This work has presented a data acquisition process resulting in a recordings collection of multimodal and multiparty interactions during a reading comprehension task in an educational setup, which has been investigated in its multimodal components and communicative functions. The data were analysed to investigate the structure of natural interaction, the multimodal conversation strategies of the participants, together with some basic behavioral patterns.

The data analysis accounts for preliminary knowledge that needs to be expanded in terms both of sample/corpus size and of the type of features investigated, i.e. to include high-level features that may be inferred from patterns and timed sequences of observed features; it provides, however, important insights towards the modelling of the participants' multimodal behavior and their multi-party interaction. Participants were bounded by certain restrictions as to the role they assumed and the task they were asked to perform. However, children's answers, the extent of the discussion during the answers phases as well as the turn management choices that children made, were as spontaneous as possible. The context restrictions, which comprise the task-dependent role play, as well as the physical and technical design, are expected to benefit the task modelling at a later stage; this set-up elicits interactional responses that may be simulated, i.e. when this task shifts to concrete HCI applications for individual or group tuition, as well as assisting children in their homework. Specifically, the target is to integrate this task modelling into the implementation of context-aware and adaptive dialogue systems able to grasp and infer individual as well as multi-party effects, and at the same time generate rich, socially believable behavior.

Both the data acquisition process and our preliminary findings provide valuable feedback in defining the design principles, the features to be taken into account and the technical requirements. They also advance the knowledge on possible ways of inferring high level conversational features such as dominance, attention and engagement.

The forms that feedback and turn management may take in this multimodal and multiparty setting provide significant insights to the design of an embodied system able to deal with the dynamics of the interaction and the conversational strategies of the participants. The conversational actions that the participants employ should be co-examined with their timing and duration to secure more accurate representations of the conversational flow. In the current analysis, special emphasis was also placed on the modality of gaze, which, in combination with head movements, have been considered useful cues for attention estimation and addressee identification [24, 25].

The current study is based on limited amounts of data, which impede further correlalations and generalisations. However, it provides initial indicators of measurable features that account for the interactional behavior and pave the way for further investigation. To address this weakness, larger data collections are definitely needed to study in detail and reliably the features of interest. As regards the preparatory phases of the experimentation, it is desirable that further factors accounting for the subjects' profile are taken into account, apart from their age and their school level, such as gender, prior knowledge of the subject, personality traits and post-assessment of experience. Such features will enable a more targeted pairing of participants as well as experimenting with different conditions.

Signals that have proved to be important to modelling sensory data also specify the use of distinct technologies to be integrated, including voice activity detection and visual focus of attention, to name a few. Such technologies can be fed indirectly in powerful frameworks defining the flow of the interaction and allowing for the incorporation of a series of modules while facilitating the communication between them (cf. IrisTK [26]).

From an educational-related interaction perspective, apart from the challenging issues in modelling the conversational dynamics in multimodal and multiparty interaction, another innovative aspect in the near future lies in the association of multimodal interaction features with cognitive aspects of text comprehension on the one hand, and with the level-appropriateness of the texts to be comprehended on the otherhand. This implies a) the enrichment of our dataset with input from cognitively oriented sensorial data, e.g. brain and eyes activity features related to attention and comprehension, and b) the involvement of several Natural Language Processing modules for text processing that are coupled with readability indicators. Thus, our current efforts attempt to approach the consolidation of modelling the rich area of multiparty interaction with text processing techniques to provide a valuable output in both social and educational aspects of interaction.

Acknowledgments Research leading to these results has been funded in part by the Greek General Secretariat for Research and Technology, KRIPIS Action, under Grant No. 448306 (POLYTRO-PON). The authors would like to thank all the session subjects for their kind participation in the experiments. The authors would also like to express their appreciation to the reviewers for their valuable feedback and constructive comments.

References

1. Clifford N, Steuer J, Tauber E (1994) Computers are social actors. In: Adelson B, Dumais S, Olson J (eds) CHI '94 Proceedings. of the SIGCHI conference on human factors in computing systems, Boston, April 1994. ACM Press, pp 72–78
2. Breazeal C (2003) Emotion and sociable humanoid robots. Int J Hum Comput Stud 59(1–2):119–155
3. Cohen P, Oviatt S (1995) The role of voice input for human-machine communication. Proc Natl Acad Sci 92(22):9921–9927
4. Kapoor A, Picard RW (2005) Multimodal affect recognition in learning environments. In: MULTIMEDIA'05 Proceedings of the 13th annual ACM international conference on Multimedia, Singapore, November 2005. ACM press, pp 677–682
5. Castellano G et al (2013) Towards empathic virtual and robotic tutors. In: Chad Lane H et al (eds) Artificial intelligence in education, vol 7926, Lecture notes in artificial intelligence. Springer, Heidelberg, pp 733–736
6. Robins B et al (2005) Robotic assistants in therapy and education of children with autism: can a small humanoid robot help encourage social interaction skills? Univers Access Inform Soc 4(2):105–120
7. Cassell J (2009) Embodied conversational agents. MIT Press, Cambridge
8. Rudnicky A (2005) Multimodal dialogue systems. In: Minker W, Buhler W, Dybkjaer L (eds) Spoken multimodal human-computer dialogue in mobile environments, vol 28. Text, speech and language technology. Springer, Dordrecht, pp 3–11
9. Al Moubayed S et al. (2012) Furhat: a back-projected human-like robot head for multiparty human-machine interaction. In: Esposito A et al. (eds) Cognitive behavioural systems, vol 7403. Lecture notes in computer science. Springer. Heidelberg, pp 114–130
10. Oertel C et al (2013) D64: a corpus of richly recorded conversational interaction. J Multimodal User Interfaces 7:19–28
11. Edlund J et al. (2010) Spontal: a Swedish spontaneous dialogue corpus of audio, video and motion capture. In: Calzolari et al. (eds) LREC 2010 Proceedings of the seventh conference on international language resources and evaluation, Valetta, May 2010. ELRA, pp 2992–2995
12. Paggio P et al. (2010) The NOMCO multimodal Nordic resource - goals and characteristics. In: Calzolari et al. (eds) LREC 2010 Proceedings of the seventh conference on international language resources and evaluation valetta, May 2010. ELRA, pp 2968–2973
13. Carletta J (2007) Unleashing the killer corpus: experiences in creating the multi-everything AMI meeting corpus. J Lang Resour Eval 41(2):181–190

14. Wittenburg P et al. (2006) ELAN: a professional framework for multimodality research. In: Calzolari et al. (eds) LREC 2006 Proceedings of the fifth conference on International language resources and evaluation, Genoa, May 2006. ELRA, pp 1556–1559
15. Allwood et al. (2007) The mumin coding scheme for the annotation of feedback, turn management and sequencing phenomena. Multimodal corpora for modelling human multimodal behaviour. J Lang Resour Eval 41(3–4):273–287
16. Koutsombogera M et al. (2014) The tutorbot corpus - A corpus for studying tutoring behaviour in multiparty face-to-face spoken dialogue. In: Calzolari et al. (eds) LREC 2014 Proceedings of the ninth conference on international language resources and evaluation. Reykjavik, May 2014. ELRA, pp 4196–4201
17. Sacks H, Schegloff E, Jefferson G (1974) A simplest systematics for the organization of turn-taking in conversation. Language 50:696–735
18. Duncan S (1972) Some signals and rules for taking speaking turns in conversation. J Pers Soc Psychol 23:283–292
19. Goodwin C (1980) Restarts, pauses and the achievement of mutual gaze at turn-beginning. Sociol Inq 50(3–4):272–302
20. Bohus D, Horvitz E (2010) Facilitating multiparty dialog with gaze, gesture, and speech. In: ICMI-MLMI '10 International Conference on Multimodal Interfaces and the Workshop on Machine Learning for Multimodal Interaction, Beijing, November 2010. ACM Press, p 311
21. Allwood J, Nivre J, Ahlsén E (1993) On the semantics and pragmatics of linguistic feedback. J Semant 9(1):1–29
22. Koutsombogera M, Papageorgiou H (2010) Linguistic and non-verbal cues for the induction of silent feedback. In: Esposito A et al. (eds) Development of multimodal interfaces: active listening and synchrony, vol 5967. Lecture notes in computer science. Springer, Heidelberg, pp 327–336
23. Allwood J et al (2007) The analysis of embodied communicative feedback in multimodal corpora: a prerequisite for behavior simulation. J Lang Resour Eval 41(3–4):255–272
24. Al Moubayed S, Skantze G (2012) Perception of gaze direction for situated interaction. In: Gaze-In '12 proceedings of the 4th workshop on eye gaze in intelligent human machine interaction, Santa Monica, October 2012. ACM Press, p 88
25. Johansson M, Skantze G, Gustafson J (2013) Head pose patterns in multiparty human-robot team-building interactions. In: Herrmann G et al. (eds) International conference on social robotics, Bristol, October 2013. Lecture notes in artificial intelligence, vol 8239. Springer International publishing, pp 351–360
26. Skantze G, Al Moubayed S (2012) IrisTK: a statechart-based toolkit for multi-party face-to-face interaction. In: ICMI'12 Proceedings of the 14th ACM international conference on multimodal interaction, Santa Monica, October 2012. ACM Press, pp 69–76

Chapter 11
Detecting Abnormal Behavioral Patterns in Crowd Scenarios

Hossein Mousavi, Hamed Kiani Galoogahi, Alessandro Perina
and Vittorio Murino

Abstract This Chapter presents a framework for the the task of abnormality detection in crowded scenes based on the analysis of trajectories, build up upon a novel video descriptor, called *Histogram of Oriented Tracklets*. Unlike standard approaches that employ low level motion features, e.g. optical flow, to form video descriptors, we propose to exploit mid-level features extracted from long-range motion trajectories called *tracklets*, which have been successfully applied for action modeling and video analysis. Following standard procedure, a video sequence is divided into spatio-temporal cuboids within which we collect statistics of the tracklets passing through them. Specifically, tracklets orientation and magnitude are quantized in a two-dimensional histogram which encodes the actual motion patterns in each cuboid. These histograms are then fed into machine learning models (e.g., Latent Dirichlet allocation and Support Vector Machines) to detect abnormal behaviors in video sequences. The evaluation of the proposed descriptor on different datasets, namely UCSD, BEHAVE, UMN and Violence in Crowds, yields compelling results in abnormality detection, by setting new state-of-the-art and outperforming former descriptors based on the optical flow, dense trajectories and social force models.

H. Mousavi · H.K. Galoogahi · A. Perina · V. Murino (✉)
Pattern Analysis and Computer Vision Department, Istituto Italiano
di Tecnologia, Genoa, Italy
e-mail: vittorio.murino@iit.it; vittorio.murino@univr.it

H. Mousavi
e-mail: hossein.mousavi@iit.it

H.K. Galoogahi
e-mail: hamed.kiani@iit.it

A. Perina
e-mail: alessandro.perina@iit.it

V. Murino
Department of Computer Science, University of Verona, Verona, Italy

© Springer International Publishing Switzerland 2016
A. Esposito and L.C. Jain (eds.), *Toward Robotic Socially Believable
Behaving Systems - Volume II*, Intelligent Systems Reference Library 106,
DOI 10.1007/978-3-319-31053-4_11

11.1 Introduction

The study of human behavior has become an active research topic in the areas of human-computer interaction, robot learning, user interface design, intelligent surveillance and crowd analysis. Among these fields, crowd analysis has recently attracted increasing attention in computer vision for several problems such as density estimation [7], motion detection [49], tracking [3] and crowd behavior detection [31, 32, 38, 43, 50].

Crowd behavior detection refers to identifying the behavioral patterns of individuals involved in a crowd scenario. These behavioral patterns may vary from the peaceful movement of pilgrims in Mecca to the violent acts of a riot in NYC streets for instance. It is well noted in the sociological literature that a crowd goes beyond a set of individuals who independently display their personal behavioral patterns [47]. In other words, the behavior of each individual in a crowd can be influenced by "crowd factors" (e.g., dynamics, goals, environment, events, etc.), and the individuals behave in a different way than if they were alone. This, indeed, implies that a crowd reflects high-level semantic behavior which can be exploited to model crowd motions [16].

Based on the above explanation, existing computer vision techniques to date designed for the detection of individual behavioral patterns are not suitable for modeling and detecting events in crowd scenes. This has encouraged the vision community to tailor techniques for modeling and understanding behavioral patterns in crowd scenarios. A large portion of recent works is dedicated to model and detect abnormal behaviors in video data. Existing works in the literature are basically different in terms of the type of abnormal behavior (e.g., panic [12], violence [13], escape [50]), types of features (histograms of low level features [37, 51, 55], optical flow [17, 29], trajectories [9], spatio-temporal features [6, 19], etc.), modeling frameworks and learning techniques such as Markov-like Models [17], Bayesian models [46], clustering Models [2, 11] and Social Force models [29, 53].

In this chapter, we propose the analysis of short trajectories through the introduction of a new video descriptor for detecting abnormal behaviors. In a nutshell, we describe each spatio-temporal window in the video volume using motion trajectories represented by a set of tracklets [36], which are short trajectories extracted by tracking interest points within a short time period. A key difference of our approach with most of the previous works is explained in Fig. 11.1. Standard approaches typically describe frames with *dense* descriptors like the optical flow [26] or the interaction force [29]. Then 2D or 3D patches are sampled from the video volumes, and abnormal and normal frames are classified by using a bag of words approach. In our case, however, we define spatio-temporal "cuboids" and we collect statistics on the *sparse* trajectories that intersect them. More in detail, the magnitude and orientation of such intersecting tracklets are encoded in a histogram which we called *histogram of oriented tracklets* (HOT) [30]. In this sense, our method directly provides a histogram to describe a frame and no clustering is needed to create the dictionary. Moreover, we will show that the tracklet extraction proposed in [30] is sensitive to noisy tracklets

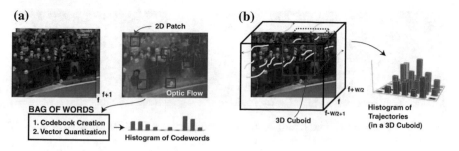

Fig. 11.1 a Standard dictionary learning based on dense motion information and 2D patches.
b Our approach based on 3D video patches and 2D histogram computing

and is not able to extract tracklets of individuals/objects appearing in the subsequent
frames of the video of interest. To tackle this drawback, we propose a robust tracklet extraction that re-initializes/detects salient points in each frame (frame level) to
detect all salient points appear within the video.

Similarly to ours, the approach proposed in [9] considered trajectories too. They,
however, exploited trajectories to compute "energy potentials" in the pixel space
describing the amount of interaction between people. Their work is more interested
in capturing "differences" between neighboring trajectories.

Under the hypothesis that abnormalities are outliers of normal situations, we
employ two standard approaches for classification: (i) when only normal data is available for training we apply the Latent Dirichlet allocation (LDA) generative model [4],
and (ii) in the cases that abnormal training data is available too we use Support Vector
Machines (SVM) discriminative learning technique for identify abnormalities.

We evaluate our approach on standard abnormality detection benchmark datasets
such as USCD [27], BEHAVE [5], Violence in Crowds [13] and the publicly available
dataset from University of Minnesota [29]. The obtained results are compared to other
descriptors based on optical flow as well as social force model [29]. Our approach
reaches very promising accuracy in frame level abnormality detection and it sets the
new state-of-the-art, while being much simpler than other techniques in the literature.

In the following section, we will detail some of leading approaches for abnormality
detection in video sequences more in detail.

11.2 Related Work

The major challenge in abnormality detection is that there is not a clear definition
of abnormality since they are basically context dependent and can be defined as outliers of normal distributions. Based on this widely accepted definition of abnormality,
existing approaches for detecting abnormal events in crowd can be generally classified into two main categories: (i) object-based approaches and (ii) holistic techniques.

Fig. 11.2 **a** Shibuya crossing (Japan). **b** Mecca (Saudi Arabia). **c** Tracklets extracted from UCSD dataset

Object-Based Approaches. Object-based methods treat a crowd as a set of different objects. The extracted objects are then tracked through the video sequence and the target behavioral patterns are inferred from their motion/interaction models (e.g. based on trajectories). This class of methods relies on the detection and tracking of people and objects. Despite promising improvements to address several crowd problems [34, 39], they are limited to low density situations and perform well when high quality videos are provided, which is not the case in real-world situations. In other words, they are not capable of handling high density crowd scenarios due to severe occlusion and clutter which make individuals/objects detecting and tracking intractable [28, 33]. Figure 11.2a, b shows an example of such scenarios. Some works made noticeable efforts to circumvent robustness issues. For instance, Zhao and Nevatia [54] used 3D human models to detect persons in the observed scene as well as a probabilistic framework for tracking extracted features from the persons. In contrast, some other methods track feature points in the scene using the well-known KLT algorithm [35, 40]. Then, trajectories are clustered using space proximity. Such a clustering step helps to obtain a one-to-one association between individuals and trajectory clusters, which is quite a strong assumption seldom verified in a crowd scenario.

Holistic Approaches. The holistic approaches, on the other hand, do not separately detect and track each individual/object in a scene. Instead, they treat the crowd as a single entity and try to exploit low/medium level features (mainly in histogram form) extracted from the video in order to analyze the sequence as a whole. Typical features used in these approaches are spatial-temporal gradients or optical flows. Krausz and Bauckhage [21, 22] employed optical flow histograms to represent the global motion in a crowded scene. The extracted histograms of the optical flow along with some simple heuristic rules were used to detect specific dangerous crowd behaviors. More advanced techniques exploited models derived from fluid dynamics or other physics laws in order to model a crowd as an ensemble of moving particles. Together with Social Force Models (SFM), it was possible to describe the behavior of a crowd by means of interaction of individuals [14]. In [29] the SFM is used to detect global anomalies and estimate local anomalies by detecting focus regions in the current frame. For abnormality detection, Solmaz and Shah [41] proposed a method to classify the critical points of a continuous dynamical system, which was applied for high-density crowds such as religious festivals and marathons [15].

In addition, several approaches deal with the complexity of a dynamic scene analysis by partitioning a given video in spatial-temporal volumes. In [19, 20] Kratz and Nishino extract spatial-temporal gradients from each pixel of a video. Then, the gradients of a spatio-temporal cell are modeled using Spatial-Temporal Motion Pattern Models, which are basically 3D Gaussian clusters of gradients. Hart and Storck [4] used group gradients observed at training time in separate cluster centers, then they used the Kullback–Leibler distance [4] in order to select the training prototype with the closest gradient distribution. Mahadevan et al. [27] model the observed movement in each spatial-temporal cell using dynamic textures, which can be seen as an extension of PCA-based representations. In each cell, all the possible dynamic textures are represented with a Mixture of Dynamic Texture (MDT) models, allowing to estimate the probability of a test patch to be anomalous. In this way, the authors show that not only temporal anomalies but also pure appearance anomalies can be detected.

Difficulty from the above methods, the major contributions of our approach are listed in the following:

1. We propose to exploit motion trajectories for the task of abnormality detection in crowd scenes. Our approach is based on tracklets, which are trajectories of salient points tracked over a short period of time. This approach represents a trade-off between object- and holistic-based approaches, since: (i) there is no need of detecting particular objects/individual in frames, and (ii) the histogram form of the proposed technique allows one to holistically capture the statistics of the tracklets (in terms of orientation and magnitudes) at both frame and video levels.

2. We propose a new descriptor called Histograms of Oriented Tracklets. The new descriptor simultaneously captures the magnitude and orientation information of a set of tracklets passing through a spatio-temporal volume by means of a bi-dimensional histogram. The integration of tracklets magnitude and orientation allows us to model complex motion patterns.

3. We introduce a simplified version of HOT which referred to as simplified-HOT, in short sHOT. This descriptor mimics the conventional histogram-of-orientations based descriptors (e.g., HOG and HOF) and shows that in the case of sparse motion information (e.g., tracklets), 2D histograms strongly outperform the standard way of summarizing statistics.

11.3 Histogram of Oriented Tracklets: A Compact Representation of Crowd Motions

Tracklets [36] are compact spatio-temporal representations of moving rigid objects. They represent fragments of an entire trajectory corresponding to the movement pattern of an individual point, generated by the frame-wise association between point localization results in the neighbor frames. Tracklets capture the evolution of patches and were originally introduced to model human motions for the task of action recognition in video sequences [36].

More formally, a tracklet is represented as a sequence of points in the spatio-temporal space as:

$$\mathbf{tr} = (p_1, \ldots p_t, \ldots p_T) \tag{11.1}$$

where each p_t represents two-dimensional coordinates (x_t, y_t) of the tth point of the tracklet in the tth frame and T indicates the length of each tracklet. Tracklets are formed by selecting regions (or points) of interest via a feature detector and by tracking them over a short period of time. Thus, one can say that tracklets represent a trade-off between optical flow and object tracking. Examples of tracklets are depicted by Fig. 11.2c.

Since different regions usually exhibit different patterns of motion, we propose a histogram based descriptor that captures the statistics of object trajectories passing through a spatio-temporal cube. The new descriptor which we named Histogram of Oriented Tracklets is clearly inspired by the recent success of histograms of features in the object recognition community, being the Histogram of Oriented Gradients (HOG) the most famous example [10]. In the following, we will elaborate the procedure of computing HOT from video sequences.

11.3.1 Tracklet Extraction

We extracted all the tracklets in a given video using standard OpenCV code.[1] Specifically, the SIFT algorithm is employed to detect possible salient points [25] in a frame, as illustrated in Fig. 11.3a. Then we employed the KLT algorithm [42], to track the salient points for T frames—Fig. 11.3b. The spatial coordinates of the tracked points are used to form the tracklets set $\mathcal{T} = \{\mathbf{tr}^n\}_{n=1}^{N}$, where N is the number of all extracted tracklets and $\mathbf{tr}^n$ refers to the nth tracklet in the video sample. The length of the tracklets T depends on the frame-rate of the sequence, the relative position of the camera and the intensity of the motion-patterns present in the scene.

We explore two different strategies for tracklet extraction, as illustrated in Fig. 11.4. In the first strategy, video-level initialization [30], tracklets are initialized using the salient points detected in the first frame of the video, and then tracked until the tracker fails. In the case of tracker failure, a re-initialization is performed to handle the tracker failure or find a new salient point. This strategy was originally introduced in [30] and its main drawbacks is that tracklets are limited to the salient points which were extracted from the first frame or the salient points detected over re-initialization process (which only happens when tracker fails). This means that new set of salient points appearing in the subsequent frames will not be fully detected, and thus, not considered for tracklet extraction. This disadvantage is not very highlighted in extracting tracklets from the public datasets, since the video clips in the datasets

---

[1] Available at http://www.ces.clemson.edu/~stb/klt/.

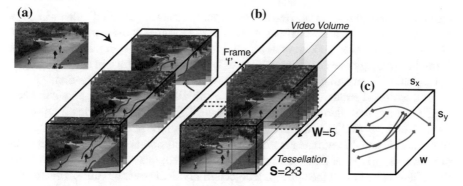

Fig. 11.3 **a** Interest points are detected and tracked over T frames to compute tracklets. **b** To compute HOT the video volume is spatially divided in non-overlapping spatial sectors (other choices are of course possible), in the figure $S = 2 \times 3$. Then for each frame we considered a temporal window stride of W frames. **c** The HOT descriptor is computed starting from the portions of tracklet in each sector of size $S_x \times S_y \times W$. In a sense *it represents the expected motion patterns in cuboids of a video*

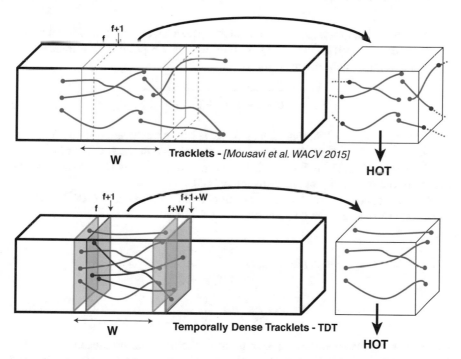

Fig. 11.4 *Top* Video level tracklet initialization. Salient points are detected at the first frame of the video and tracked over all frames. Salient point detection/re-initialization is performed when either the tracklet ends or when the tracker fails. *Bottom* frame level tracklet extraction which is called Temporally Dense Tracklets. In this strategy, salient points are re-initialized/detected at every single frame and tracked over W subsequent frames. This significantly reduces extracting noisy tracklets and is able to extract new tracklets over a given video

are short (cropped from large videos) and appearing new objects/individuals in clips are limited. In real-world situations where the videos are captured in hours, days or even weeks, however, appearing new individuals/objects is a norm. As a result, this extraction strategy does not work well to detect all salient point and consequently is not capable of extracting corresponding tracklets for abnormality detection.

In the second strategy, we proposed to re-initialize/detect salient points in each single frame of the video and track the points over T frames, we called this Temporally Dense Tracklets (TDT). This strategy is not limited to the points detected at the first frame and is capable of detecting all possible salient points over a given video. In other words, no matter how long is the captured video, this strategy is able to detect the salient points of all the appearing objects/individuals over the time. This results in producing a large pool of tracklets $\mathscr{T}$ which can used to summarize the motion-patterns observed in the scene per each frame.

11.3.2 HOT Computation

As previously mentioned, tracklets are short sequences of two-dimensional points represented as $\mathbf{tr} = \{(x_1, y_1) \cdots (x_T, y_T)\}$. For each tth point over a tracklet, the local magnitude can be computed as:

$$m_t = \sqrt{(x_{t+1} - x_t)^2 + (y_{t+1} - y_t)^2} \tag{11.2}$$

The process of HOT computation starts by splitting the video in spatio-temporal cuboids of size $S_x \times S_y \times W$ with overlapping cuboids in the spatial domain; this is illustrated by Fig. 11.3b. From now on, we will use the apex (i, s) to address the portion of the tracklet i that intersects the cuboid s.

For each cuboid s, we compute the magnitude and orientation of the portion (fragment) of each tracklet that intersects s as follows:

$$\theta^{i,s} = \arctan \frac{\left(y_{end}^{i,s} - y_{begin}^{i,s}\right)}{\left(x_{end}^{i,s} - x_{begin}^{i,s}\right)} \tag{11.3}$$

$$M^{i,s} = \max_{t \in W} \left\{ m_t^{(i,s)} \right\} \tag{11.4}$$

where m_t's are the magnitude in each point of the tracklet, introduced in Eq. 11.2. The entry and exit points of tracklet i in/from cuboid s are respectively indexed by $(x_{begin}^{i,s}, y_{begin}^{i,s})$ and $(x_{end}^{i,s}, y_{end}^{i,s})$. The process of computing the magnitude and orientation of a tracklet within a cuboid is illustrated in Fig. 11.5a, b.

Finally, the magnitudes and orientations of all tracklets passing across a cuboid are independently quantized in O orientations and M magnitudes bins. We populated the bins of a histogram $H_{\theta,m}^s$ by simply counting how many times we observe a partic-

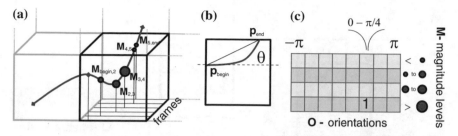

Fig. 11.5 The process of HOT formation. **a** The magnitude of the tracklet, here illustrated by red circles, is computed in every point (frame) of the temporal window. The maximum magnitude is considered, $m_{3,4}$. **b** To compute the orientation only the entry and exit point from the sector are considered. **c** Each tracklet gives a contribution to the HOT histogram, in this case we have $O = 8$ orientations and $M = 4$ magnitude bins

Fig. 11.6 On the *top* few frames for ped1 and ped2. On the *bottom*, the HOT descriptor averaged over the orientations and projected in each sector on the image plane, showing the expected motion in each sub-window

ular magnitude-orientation pair $\{\theta, m\}$, Fig. 11.5c. We normalized the histograms to make the histogram representation independent of the quantity of motions observed. This procedure computes a two-dimensional histogram called Histogram of Oriented Tracklets.

To compute a descriptor at the frame level, $H_{\theta,m}^{s,f}$, we employed a sliding window approach, where the histogram at frame f is based on cuboids that temporally ranges from $f - \frac{W}{2}$ to $f + \frac{W}{2}$. The major drawback of this choice is that abnormalities are detected with some latency due to (i) the need of the "future" frames and (ii) the fact that the descriptor of the frame when abnormality begins still encompasses information from normal "past" frames. However, this results in at most 0.2–0.5 s of latency for typical values used in this work, thus acceptable. Moreover, real systems would probably consider as outliers abnormal events of less than a second.

Summarizing, the HOT descriptor represents the expected motion patterns that happen in each sub-regions. It encompasses the magnitude of a motion and its orientation, being the latter especially useful in single camera scenarios, and often disregarded by the state of the art. Figure 11.6 shows the projection on the image plane of the HOT descriptors for 6 frames from the UCSD dataset, the magnitude is shown with a white intensity. Details on the choice of parameters will be given in the experimental section.

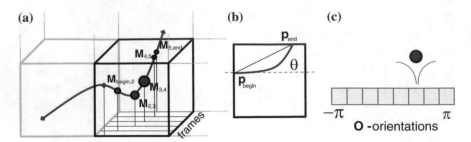

Fig. 11.7 The process of sHOT formation. **a** The magnitude of the tracklet, here illustrated by red circles, is computed in every point (frame) of the temporal window. The maximum magnitude is considered, $M_{3,4}$. **b** To compute the orientation only the entry and exit point from the sector are considered. **c** Differently from HOT, now each tracklet gives an a contribution proportional to its magnitude in the appropriate orientation bin

sHOT Descriptor: In a different approach, following the standard procedure of histogram of oriented gradients (HOG) computation [10], we propose the simplified-HOT descriptors (sHOT) which is the one-dimensional version of HOT, Fig. 11.7.

Given a set of magnitude-orientation pairs $\{\theta^n, M^n\}$, a sHOT descriptor is computed by accumulating the magnitudes whose corresponding orientations fall in orientation bins. This process is followed by a normalization to form a non-biased oriented histogram. Similar to HOT, sHOT is computed for each cuboid s temporally centered at each frame f, $h_{\theta,m}^{s,f}$.

11.4 Abnormality Detection

Unlike the usual classification tasks in computer vision, for crowd abnormality detection we hardly can assume that abnormal footage is available at the training step. This would be in fact, hard and costly to collect and it would only represent a partial view of what an abnormality is, unless we restrict to a particular scenario such as violence or panic. On the other hand, we can assume to have plenty of normal data.

The common choice to formalize mathematically "what a concept is" (in this case a normal behavior), are generative models. Generative models encode the process of data generation by representing a signal by means of a joint probability distribution. When a new observation arrives, adequately trained the generative model can assign a probability, or likelihood, that has been generated by the modeled process.

In the context of abnormal behavior detection, previous works employed latent Dirichlet allocation—LDA [4] or mixture models [29]. In this framework, we limit our attention to the former. LDA defines what is normal in terms of co-occurrences between features which are the motion patterns in our case.

Given a set of two dimensional histograms $H_{\theta,m}^{s,f}$ (or $h_{\theta,m}^{s,f}$ in the simplified version), for each frame $f = 1, \ldots, F$, we construct the LDA training corpus $\mathscr{D}$ based on two different detection strategies:

Fully bag of words—BW. In this first case, HOT descriptors are summed across spatial sectors, disregarding the spatial information:

$$D^f = \sum_s H_{\theta,m}^{s,f} \quad \text{and} \quad \mathscr{D} = \{D^f\}_{f=1}^F \qquad (11.5)$$

This strategy is useful when training and test data come from different environments [13, 29], or when a large perspective distortion is present in a video.

Per-frame, Per-sector—FS. In the second case, HOTs from all the different sectors are concatenated in a single descriptor to preserve the spatial information of each frame:

$$D^f = \{H_{\theta,m}^{1,f}|H_{\theta,m}^{2,f}|\ldots|H_{\theta,m}^{S,f}\} \quad \text{and} \quad \mathscr{D} = \{D^f\}_{f=1}^F \qquad (11.6)$$

In this case, LDA captures correlations between motion patterns that occur in different sectors of the scene.

Once $\mathscr{D}$ is collected, we employ LDA to learn a set of Z topics that defines what normality is. We used the variational Expectation Maximization algorithm (EM) which iteratively maximizes a bound on the data log-likelihood

$$\mathscr{L}(\mathscr{D}|\alpha, \beta) = \sum_f \log p(\mathscr{D}^f|\alpha, \beta) \qquad (11.7)$$

where α is the Dirichlet prior over topic combinations, and β encodes the motion patterns associated to each topic (and eventually each sector).

Using a learned model, we can estimate the log-likelihood of an unseen test frame $\mathscr{L}(\mathscr{D}_{unseen}^f|\hat{\alpha}, \hat{\beta})$ and assign either normal or abnormal label to the new frame based on a fixed threshold on the estimated likelihoods. When testing data of normal and abnormal events is available, SVM is used in a standard way (see Sect. 11.5).

11.5 Experimental Evaluation

We compare the HOT descriptor and its variations with state-of-the-art methods and descriptors in the literature, mainly the mixtures of dynamic textures framework [27] and leading optical flows based approaches [1, 9, 13, 18].

11.5.1 Crowd Datasets

Four publicly available datasets are employed for the evaluation, including USCD [27], UMN [29], Violence-In-Crowds [13] and BEHAVE [5].

UCSD Dataset[2] consists of videos of a crowded pedestrian walkway with manually collected frame-level ground truth. The dataset contains two smaller subsets corresponded to two different scenes. The first, denoted by "ped1" contains clips of 158 × 238 pixels, which depict groups of people walking toward and away from the camera, with a certain degree of perspective distortion. The second, denoted by "ped2" has a spatial resolution of 240 × 360 pixels and depicts a scene where most pedestrians move horizontally. The video footage of each scene is sliced into clips of 120–200 frames. A number of these (34 in Ped1 and 16 in Ped2) are to be used as training set for the condition of normalcy. The test set contains clips (36 for Ped1 and 12 for Ped2) with both normal (around 5,500) and abnormal (around 3,400) frames (see Fig. 11.8a). We only considered anomaly at the frame level for this dataset.

BEHAVE Dataset[3] consists of a set of complex group activities including *meeting, splitting up, standing, walking together, ignoring each other, escaping, fighting* and *running*. Following [9], the *fighting* activity is selected as abnormalities (50 clips) and the rest as normal activities (271 clips), see Fig. 11.8b.

UMN Dataset[4] includes 11 different scenarios of a panic and normal situations in three different indoor and outdoor scenes. Figure 11.8c shows some samples from UMN dataset. **Violence in Crowds Dataset**[5] (Fig. 11.8d) is composed by real-world, video footage of crowd violence, along with standard benchmark protocols designed to test violent- and non-violent classification. It is split into five sets: half violent crowd behavior and half non-violent behavior which are available at training time. It gives us the chance to test HOT descriptor with standard SVM.

11.5.2 Tracklet Extraction

As mentioned earlier, we modified the tracklet extraction method used in [30] by initializing the KLT tracker at each frame (temporally dense tracklets). More precisely, we reset interest points in each frame when track it over T frames. In video-based tracklet extraction, on the other hand, we stick to an initial set of points detected at the first frame of the video and track them for the entire length of the tracklet, re-initializing only in case of failures.

In general, we observed that the former works much better qualitatively and quantitatively. The reason is probably due to removing wrong/noisy points and capturing salient points which appear at subsequent frames. We employed the modified version

---

[2] Available at http://www.svcl.ucsd.edu/projects/anomaly/.

[3] Available at http://groups.inf.ed.ac.uk/vision/behavedata/interactoins/.

[4] http://mha.cs.umn.edu/movies/crowdactivity-all.avi.

[5] Available at http://www.openu.ac.il/home/hassner/data/violentflows/.

Fig. 11.8 Normal and abnormal samples selected from **a** UCSD dataset, **b** BEHAVE dataset, **c** UMN dataset. Here abnormality is panic with people running away. **d** Normal and Violent crowds from Violence-in-crowds dataset. Videos are from different scenes

for all the four datasets, and specifically on UCSD, we compare it with the tracklet extraction of [30].

11.5.3 HOT Parameters and Setting

Like other descriptors, there are few parameters and constants to tweak, some of them are peculiar of the descriptor, others depend on the scene monitored. These parameters are illustrated in Fig. 11.9, including spatial tessellation of the frame S, temporal window W, length of tracklets T and the quantization bins O and M. Here, instead of proposing a "gold standard" and reporting the best result obtained, we present the results varying the key parameters.

Following previous works, we quantized tracklets orientation in $O = 8$ uniform bins [30]. Unlike [30], we varied the temporal window of $W = \{5, 11, 21\}$ frames setting the tracklet length to W. Moreover, we varied the number of quantization levels for magnitude $M \in \{3, 5, 8, 16, 24, 32\}$. Quantization intervals are generated using linear spacing (i.e., Matlab's linspace) starting from 0 up to $C \times m_{max}$ where m_{max} is the maximum value of magnitude found in the training set. Abnormal data is usually not available at training time and may present quite different magnitudes ranges, therefore this range extension is necessary. We set $C = 2$, but it is worth noting that C depends on the scene analyzed and on the motion expected. Furthermore, in

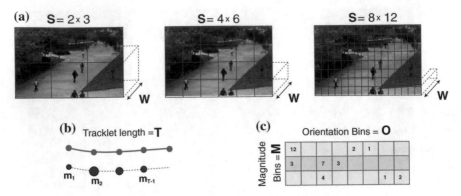

Fig. 11.9 Parameters to consider in computing the histogram of oriented tracklets: tracklet length (T), spatial tessellation (S), temporal stride (W), orientation bins (O) and magnitude bins (M)

the case of fixed camera scenarios, the maximum magnitude of a tracklets is related to the maximum speed and it can be easily bounded manually during the system set-up. In experiments, varying C did not change the result up to reasonable values. If abnormal data is available at train time, [13], C can be set to 1. Finally, we considered three different spatial tessellations, called as *coarse* $S = 2 \times 3$, *medium* $S = 4 \times 6$ and *fine* $S = 8 \times 12$.

Finally, unlike [30] which reported the best results across various choices of LDA topics Z, here we instead fixed this number to $Z = 30$ to have a much fairer comparison. Although experimentally we found that, likewise [30], the performance does not vary much changing the number of LDA topics.

Parameters Analysis. The result of parameters tuning is shown in Figs. 11.10, 11.11 and 11.12, reporting Equal Error Rate (EER)[6] as a function of magnitude levels (M), tracklet length (W) and spatial tessellation S. This experiment was conducted on the UCSD dataset, ped1 and ped2, comparing our approach with the method proposed in [30] using two different classification sensations: *Fully bag of words (BW)* and *Per-frame, Per-sector (FS)*. For our method, we considered both HOT computations: 2D HOT with Temporally Dense Trajectory (TDT) and 1D simplified HOT (sHOT). Please note that sHOT results are independent of M, therefore, it's EER in each diagram does not change regarding different magnitude levels M.

According to Figs. 11.10, 11.11 and 11.12, the highest EER is obtained by sHOT for almost all parameter combinations. Moreover, the proposed tracklet extraction (TDT) outperformed the video level initialization employed by original HOT [30], especially in ped2 subset. This is caused by the disadvantage of HOT's tracklet extraction to re-track salient points affected by occlusion or detect new salient points over subsequent frames. Further, in general, fully bag of words (BW) detection strategy achieved slightly better detection performance than Per-frame, Per-sector (FS).

---

[6]The Equal Error Rate is the value of false positive rate when the ROC curve intersects the line connecting $(0, 1)$–$(1, 0)$.

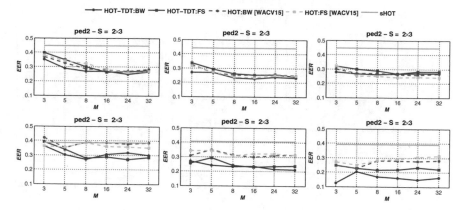

Fig. 11.10 Results for `ped1` and `ped2` varying the number of magnitude bins and tracklet length for the coarse spatial tessellation

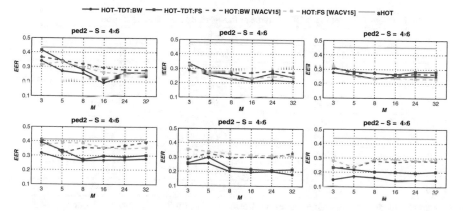

Fig. 11.11 Results for `ped1` and `ped2` varying the number of magnitude bins and tracklet length for the medium spatial tessellation

This states that holistic motion statistics captured by BW are more discriminative than those computed locally by FS. By increasing the level of tessellation from fine to coarse (more holistic) the difference between EER of BW and FS decreases. Inspecting the reported results, one can see that the EER decreases by increasing the number of magnitude levels. The EER obtained by tracklet length $W = 11$ (for `ped1`) and $W = 21$ (for `ped2`) are slightly less than the tracklet length of 5. The best results obtained by our technique, the original HOT [30] and some leading approaches [1, 24, 27, 29] are represented in Table 11.1.

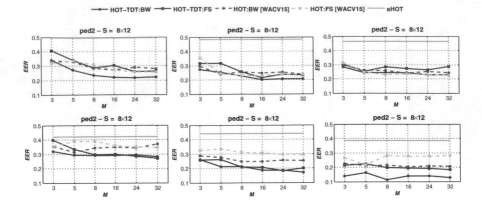

Fig. 11.12 Results for ped1 and ped2 varying the number of magnitude bins and tracklet length for the fine spatial tessellation

Table 11.1 Equal error rates on UCSD dataset using standard testing protocol

ped1		ped2	
Method	EER (%)	Method	EER (%)
MDT [24]	22.90	MDT [24]	27.90
SFM [29]	36.50	SFM [29]	35.00
LMH [1]	38.90	LMH [1]	45.80
HOT: BW [30]	23.84	HOT: BW [30]	20.42
HOT: FS [30]	22.53	HOT: FS [30]	21.84
HOT: BW-TDT	**19.09**	HOT: BW-TDT	**11.24**
HOT: FS-TDT	**21.09**	HOT: FS-TDT	**18.15**
sHOT	43.03	sHOT	38.45

The results of the previous approaches are borrowed from [27]. The results of our approach and the original HOT are the best performance obtained at the parameter tuning experiment, shown in Figs. 11.10, 11.11 and 11.12. The LDA topics, Z, is fixed to 30 for our approach and the original HOT [30]

11.5.4 Detection Performance

In this following, we evaluate the detection performance of our approach comparing with the state-of-the-arts. For UCSD dataset, we reported the best performance (smallest EER) achieved at the previous experiment respecting to different settings of the parameters. For the other datasets- UMN, Violent in crowd and BEHAVE- the parameters are fixed to $W = 11$, $M = 16$ and $O = 8$. The classification strategy for the original HOT and the proposed approach is limited to fully bag of words (BW).

Evaluation on UCSD dataset. For the UCSD dataset we considered its standard train-test partition following [30]. The LDA likelihood [4] of the test frames was used to compute the EERs for our approach and the original HOT [30]. Results (EER,

Table 11.2   AUC on UMN dataset, comparing our approach with the state-of-the-arts

Dataset	HOT-TDT	sHOT	HOT	SFM	SR	OF	CI
Scene-1	**0.998**	**0.998**	0.993	0.990	0.995	0.964	n/a
Scene-2	0.991	**0.995**	0.984	0.949	0.975	0.906	n/a
Scene-3	**0.998**	0.996	0.991	0.989	0.964	0.967	n/a
All scenes	**0.994**	0.984	0.991	0.960	0.978	0.840	0.990

the smaller the better) for `ped1` and `ped2` are reported in Table 11.1. The results of our approach (HOT:FS-TDT, HOT:BW-TDT and sHOT) are the lowest EER of each strategy across all the parameters. EERs of competitors are taken from [27] where the authors reported *best* results across all the model-method configurations.

Despite such comparison cannot statistically highlight a clear winner, we limit ourselves to acknowledge how the new tracklet extraction strategy (TDT) proposed here improves the performance of [30] and, surely, outperforms the prior leading methods in the literature. Particularly, our approach achieved superior performance than the original HOT for both classification strategies, BW and FS, on `ped1` and `ped2`. The difference between the performance of HOT·BW·TDT on `ped1` and `ped2` (19.09 vs. 11.24 %) can be rationally explained with the presence of perspective distortion in `ped1` which may affect extracting discriminative tracklets. This problem may be solved by some rough estimation of scene geometry which is out of the scope of this work. Please note that the EERs of the original HOT reported in Table 11.1 are slightly different than those obtained in the reference paper [30]. Since, here we fixed the LDA topics $Z = 30$, while, the results in [30] were the best achieved by varying $Z \in \{2, 4, 6, \ldots, 80\}$.

Evaluation on UMN dataset. In this experiment, we compared the proposed approach with HOT [30], social force model (SFM) [29], sparse reconstruction (SP) [8], optical flow (OF) [29], Chaotic Invariants (CI) [48] following the standard evaluation of [29]. To have a finer evaluation, we deployed a protocol by consideration of UMN three scenes separately. We found this protocol so helpful to analyses the effect of proposed descriptor in each scene individually. The results on each scene (*scene-1*, *scene-2*, *scene-3*) and the whole dataset (*all scenes*) are reported in Table 11.2 in terms of AUC (Area Under the ROC Curve). The result demonstrates the superiority of our approach on this dataset for both scene-based and all scenes evaluations.

Evaluation on Violence in Crowds dataset. In this experiment, we trained SVM with linear kernel on a set of normal and abnormal videos across a five-fold cross validation. Unlike frame classification in the previous experiments, here, the goal is

Table 11.3 Classification
results on violence in crowds
dataset

Method	Accuracy (%)
Local trinary patterns [52]	71.53
Histogram of oriented gradients [23]	57.43
Histogram of oriented optic-flow [45]	58.53
HNF [23]	56.52
Violence flows ViF [13]	81.30
Dense trajectories [44]	78.21
HOT:BW [30]	82.30
HOT:BW-TDT	**82.84**

to assign either a normal or abnormal label to an input video clip (video level classification). The video level descriptor D^V of an input video V is simply computed by:

$$D^V = \sum_{f \in V} D^f \qquad (11.8)$$

where D^f for BW is defined in Eq. 11.5. Finally, the training sets to train the SVM are formed as $\mathscr{D} = \{D^V\}_{V=1}^N$, where N is the number of positive and negative training videos. The result of this evaluation is reported in Table 11.3 showing that the best performance belongs to our approach (82.84%) followed by the original HOT (82.30%). The descriptors in [23, 44, 52] were originally proposed for action recognition in videos. But, these descriptors (except dense trajectories) were evaluated for abnormal detection in [13], where we borrowed results form. We evaluated dense trajectories [44] to complete our comparison.

Evaluation on Behave dataset. This experiment compares our method with the optical flow based method, social force model and interaction energy potential [9]. Following settings in [9], we used half of normal and abnormal videos for training and the rest for testing. We used BW classification strategy along with linear kernel SVM for frame level classification. The positive and negative training descriptors $\mathscr{D}$ formed by Eq. 11.5. The results are reported by the means of ROC as shown in Fig. 11.13.

11.6 Conclusions

In this Chapter, we introduced the Histogram of Oriented Tracklets (HOT) descriptor for the task of abnormality detection in crowd scenes. This new descriptor entails both orientation and magnitude statistics in a single feature, a situation that is often reached by combining multiple descriptors. Moreover, we empirically showed that tracklet

Fig. 11.13 ROC curve on
BEHAVE dataset

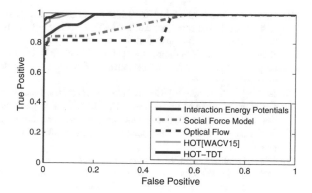

extraction based on video level salient points initialization does not perform well in crowded scenes, which can drastically degrade the performance of abnormality detection. To tackle this drawback, we proposed a variant of the HOT descriptor consisting in re-initializing salient points at each frame in the temporal domain. Experimental results showed how the analysis of trajectories can effectively capture the anomalies in crowd scenes, by setting new state of the art results on benchmark datasets.

References

1. Adam A, Rivlin E, Shimshoni I, Reinitz D (2008) Robust real-time unusual event detection using multiple fixed-location monitors. IEEE Trans Pattern Anal Mach Intell 30:555–560
2. Alvar M, Torsello A, Sanchez-Miralles A, Armingol JM (2014) Abnormal behavior detection using dominant sets. Mach Vis Appl 32(6):1–18
3. Bera A, Manocha D (2014) Realtime multilevel crowd tracking using reciprocal velocity obstacles. arXiv:1402.2826
4. Blei DM, Ng AY, Jordan MI (2003) Latent dirichlet allocation. J Mach Learn Res 3:993–1022
5. Blunsden S, Fisher R (2010) The behave video dataset: ground truthed video for multi-person behavior classification. Ann Br Mach Vis Assoc 4:1–12
6. Boiman O, Irani M (2007) Detecting irregularities in images and in video. Int J Comput Vis 74:17–31
7. Chen K, Gong S, Xiang T, Loy CC (2013) Cumulative attribute space for age and crowd density estimation. In: Proceedings of the IEEE conference on computer vision and pattern recognition, pp 2467–2474
8. Cong Y, Yuan J, Liu J (2011) Sparse reconstruction cost for abnormal event detection. In: Proceedings of the IEEE conference on computer vision and pattern recognition, pp 3449–3456
9. Cui X, Liu Q, Gao M, Metaxas DN (2011) Abnormal detection using interaction energy potentials. In: Proceedings of the IEEE conference on computer vision and pattern recognition, pp 3161–3167
10. Dalal N, Triggs B (2005) Histograms of oriented gradients for human detection. In: Proceedings of the IEEE conference on computer vision and pattern recognition, vol 1, pp 886–893
11. Fu Z, Hu W, Tan T (2005) Similarity based vehicle trajectory clustering and anomaly detection. In: Proceedings of the IEEE international conference on image processing, vol 2, pp II–602

12. Haque M, Murshed M (2010) Panic-driven event detection from surveillance video stream without track and motion features. In: Proceedings of the IEEE international conference on multimedia and expo, pp 173–178
13. Hassner T, Itcher Y, Kliper-Gross O (2012) Violent flows: real-time detection of violent crowd behavior. In: Proceedings of the IEEE conference on computer vision and pattern recognition workshops, pp 1–6
14. Helbing D, Molnar P (1995) Social force model for pedestrian dynamics. Phys Rev E 51:4282
15. Hu M, Ali S, Shah M (2008) Detecting global motion patterns in complex videos. In: Proceedings of the international conference on pattern recognition, pp 1–5
16. Junior SJ et al (2010) Crowd analysis using computer vision techniques. IEEE Signal Process Mag 27:66–77
17. Kim J, Grauman K (2009) Observe locally, infer globally: a space-time mrf for detecting abnormal activities with incremental updates. In: Proceedings of the IEEE conference on computer vision and pattern recognition, pp 2921–2928
18. Kim J, Grauman K (2009) Observe locally, infer globally: a space-time MRF for detecting abnormal activities with incremental updates. In: Proceedings of the IEEE conference on computer vision and pattern recognition, pp 2921–2928
19. Kratz L, Nishino K (2009) Anomaly detection in extremely crowded scenes using spatio-temporal motion pattern models. In: Proceedings of the IEEE conference on computer vision and pattern recognition, pp 1446–1453
20. Kratz L, Nishino K (2010) Tracking with local spatio-temporal motion patterns in extremely crowded scenes. In: Proceedings of the IEEE conference on computer vision and pattern recognition, pp 693–700
21. Krausz B, Bauckhage C (2011) Analyzing pedestrian behavior in crowds for automatic detection of congestions. In: Proceedings of the IEEE international conference on computer vision workshops, pp 144–149
22. Krausz B, Bauckhage C (2012) Loveparade 2010: automatic video analysis of a crowd disaster. Comput Vis Image Underst 116(3):307–319
23. Laptev I, Marszalek M, Schmid C, Rozenfeld B (2008) Learning realistic human actions from movies. In: Proceedings of the IEEE conference on computer vision and pattern recognition
24. Li W, Mahadevan V, Vasconcelos N (2014) Anomaly detection and localization in crowded scenes. IEEE Trans Pattern Anal Mach Intell 36:18–32
25. Lowe DG (2004) Distinctive image features from scale-invariant keypoints. Int J Comput Vis 60:91–110
26. Lucas BD, Kanade T et al (1981) An iterative image registration technique with an application to stereo vision. In: IJCAI, vol 81, pp 674–679
27. Mahadevan V, Li W, Bhalodia V, Vasconcelos N (2010) Anomaly detection in crowded scenes. In: Proceedings of the IEEE conference on computer vision and pattern recognition, pp 1975–1981
28. Marques JS, Jorge PM, Abrantes AJ, Lemos J (2003) Tracking groups of pedestrians in video sequences. In: Proceedings of the IEEE conference on computer vision and pattern recognition workshop, vol 9, pp 101–101
29. Mehran R, Oyama A, Shah M (2009) Abnormal crowd behavior detection using social force model. In: Proceedings of the IEEE conference on computer vision and pattern recognition, pp 935–942
30. Mousavi H, Mohammadi S, Perina A, Chellali R, Murino V (2015) Analyzing tracklets for the detection of abnormal crowd behavior. In: Proceedings of the IEEE Winter conference on applications of computer vision, pp 148–155
31. Mousavi H, Nabi M, Kiani H, Perina A, Murino V (2015) Crowd motion monitoring using tracklet-based commotion measure. In: Proceedings of the IEEE international conference on image processing
32. Mousavi H, Nabi M, Kiani Galoghahi H, Perina A, Murino V (2015) Abnormality detection with improved histogram of oriented tracklets. In: Proceedings of the international conference on image analysis and processing

33. Piciarelli C, Micheloni C, Foresti GL (2008) Trajectory-based anomalous event detection. IEEE Trans Circuits Syst Video Technol 18:1544–1554
34. Rabaud V, Belongie S (2006) Counting crowded moving objects. In: Proceedings of the IEEE conference on computer vision and pattern recognition
35. Rabaud V, Belongie S (2006) Counting crowded moving objects. In: Proceedings of the IEEE conference on computer vision and pattern recognition, pp 705–711
36. Raptis M, Soatto S (2010) Tracklet descriptors for action modeling and video analysis. In: Proceedings of the IEEE European conference on computer vision
37. Reddy V, Sanderson C, Lovell BC (2011) Improved anomaly detection in crowded scenes via cell-based analysis of foreground speed, size and texture. In: Proceedings of the IEEE conference on computer vision and pattern recognition workshops
38. Regazzoni C, Tesei A, Murino V (1993) A real-time vision system for crowding monitoring. In: Proceedings of the IECON'93, International conference on industrial electronics, control, and instrumentation, pp 1860–1864
39. Rittscher J, Tu PH, Krahnstoever N (2005) Simultaneous estimation of segmentation and shape. In: CVPR, vol 2, pp 486–493. IEEE
40. Shi J, Tomasi C (1994) Good features to track. In: Proceedings of the IEEE conference on computer vision and pattern recognition, pp 593–600
41. Solmaz B, Moore BE, Shah M (2012) Identifying behaviors in crowd scenes using stability analysis for dynamical systems. IEEE Trans Pattern Anal Mach Intell 34:2064–2070
42. Tomasi C, Kanade T (1991) Detection and tracking of point features
43. Wang B, Ye M, Li X, Zhao F, Ding J (2012) Abnormal crowd behavior detection using high-frequency and spatio-temporal features. Mach Vis Appl 23:501–511
44. Wang H, Kläser A, Schmid C, Liu C-L (2011) Action recognition by dense trajectories. In: Proceedings of the IEEE conference on computer vision and pattern recognition
45. Wang T, Snoussi H (2012) Histograms of optical flow orientation for visual abnormal events detection. In: AVSS, pp 13–18
46. Wang X, Ma X, Grimson WEL (2009) Unsupervised activity perception in crowded and complicated scenes using hierarchical bayesian models. IEEE Trans Pattern Anal Mach Intell 31:539–555
47. Wijermans N, Jorna R, Jager W, Vliet TV (2007) Modelling crowd dynamics, influence factors related to the probability of a riot. In: Proceedings of the fourth European social simualtion association conference (ESSA). Toulouse University of Social Sciences
48. Wu S, Moore BE, Shah M (2010) Chaotic invariants of lagrangian particle trajectories for anomaly detection in crowded scenes. In: Proceedings of the IEEE conference on computer vision and pattern recognition, pp 2054–2060
49. Wu S, San Wong H (2012) Crowd motion partitioning in a scattered motion field. IEEE Trans Syst Man Cybern Part B: Cybern 42:1443–1454
50. Wu S, Wong H-S, Yu Z (2014) A Bayesian model for crowd escape behavior detection. IEEE Trans Circuits Syst Video Technol 24(1):85–98
51. Xiang T, Gong S (2008) Video behavior profiling for anomaly detection. IEEE Trans Pattern Anal Mach Intell 30:893–908
52. Yeffet L, Wolf L (2009) Local trinary patterns for human action recognition. In: Proceedings of the IEEE international conference on computer vision
53. Zhang Y, Qin L, Ji R, Yao H, Huang Q (2014) Social attribute-aware force model: exploiting richness of interaction for abnormal crowd detection
54. Zhao T, Nevatia R (2003) Bayesian human segmentation in crowded situations. In: Proceedings of the IEEE conference on computer vision and pattern recognition, vol 2, pp II–459
55. Zhong H, Shi J, Visontai M (2004) Detecting unusual activity in video. In: Proceedings of the IEEE conference on computer vision and pattern recognition, vol 2

Author Index

© Springer International Publishing Switzerland 2016
A. Esposito and L.C. Jain (eds.), *Toward Robotic Socially Believable Behaving Systems - Volume II*, Intelligent Systems Reference Library 106, DOI 10.1007/978-3-319-31053-4

Printed in the United States
By Bookmasters